RECONFIGURING NATURE

Cardiff Papers in Qualitative Research

About the Series

The Cardiff School of Social Sciences at Cardiff University is well known for the breadth and quality of its empirical research in various major areas of sociology and social policy. In particular, it enjoys an international reputation for research using qualitative methodology, including qualitative approaches to data collection and analysis.

This series publishes original sociological research that reflects the tradition of qualitative and ethnographic inquiry developed at Cardiff in recent years. The series includes monographs reporting on empirical research, collections of papers reporting research on particular themes and other monographs or edited collections on methodological developments and issues.

Reconfiguring Nature

Issues and Debates in the New Genetics

Edited by

PETER GLASNER
Cardiff Universtiy

ASHGATE

Published by
Ashgate Publishing Limited
Gower House
Croft Road
Aldershot
Hampshire GU11 3HR
England

Ashgate Publishing Company
Suite 420
101 Cherry Street
Burlington, VT 05401-4405
USA

Ashgate website: http://www.ashgate.com

British Library Cataloguing in Publication Data
Reconfiguring nature : issues and debates in the new genetics. - (Cardiff papers in qualitative research)
1. Human genetics - History 2. Human genetics - Moral and ethical aspects 3. Human genetics - Social aspects 4. Genetic engineering - Social aspects 5. Biotechnology - Social aspects
I. Glasner, Peter E. (Peter Egon)
599.9'35

Library of Congress Cataloging-in-Publication Data
Reconfiguring nature : issues and debates in the new genetics / edited by Peter Glasner.
p. cm. -- (Cardiff papers in qualitative research)
Includes bibliographical references and index.
ISBN 0-7546-3237-7
1. Genetics--Social aspects. 2. Qualitative research. I. Glasner, Peter E. II. Series.

QH438.7.R43 2003
303.48'3--dc21 2003052116

ISBN 0 7546 3237 7

Printed in Great Britain by
Antony Rowe Ltd, Chippenham, Wiltshire

Contents

PART III: GEN-ETHICS: HUMAN GENETIC BANKING

PART IV: GENETIC SCREENING

PART V: CLONING AND XENOTRANSPLANTATION

List of Tables and Figures

Tables

Figures

List of Contributors

Professor Kåre Berg, Institute of Medical Genetics, University of Oslo, Norway.

Dr Susanne Braun, University of Luneberg, Germany.

Dr Nik Brown, Science and Technology Studies Unit (SATSU), University of York.

Professor Sarah Franklin, ESRC Centre for the Study of the Economic and Social Aspects of Genomics (CESAGen), University of Lancaster.

Professor Peter Glasner, ESRC Centre for the Study of the Economic and Social Aspects of Genomics (CESAGen), Cardiff University.

Dr Jürgen Hampel, Centre for Technology Assessment in Baden-Wurttenburg, Stuttgart.

Dr Adam M. Hedgecoe, School of Social Sciences, University of Sussex.

Professor Alan Irwin, Faculty of Human Sciences, Brunel University.

Jane Kaye, Faculty of Law, St. Catherine's College, University of Oxford.

Dr Veikko Launis, Department of Philosophy, University of Turku, Finland.

Dr Mairi Levitt, Centre for Professional Ethics, University of Central Lancashire.

Dr Alun McCarthy, GlaxoWellcome, Greenford, Middlesex.

Dr Paul Martin, Genetics and Society Programme, IGBIS, University of Nottingham.

Dr Annemiek Nelis, University of Amsterdam, Holland.

Professor Harry Rothman, Institute for Enterprise and Innovation, Nottingham University Business School.

Dr Doris Schroeder, European Academy for the Study of the Consequences of Scientific and Technological Advances, Germany.

Dr Juergen Simon, University of Luneberg, Germany.

Dr Dirk Stemerding, School of Philosophy and Social Sciences, University of Twente, Holland.

Dr Jill Turner, School of Nursing and Midwifery, Queen's University, Belfast.

Dr John Turney, Science and Technology Studies, University College, London.

Dr Garrath Williams, ESRC Centre for the Study of the Economic and Social Aspects of Genomics (CESAGen), Lancaster University.

Acknowledgements

I would like to acknowledge the Economic and Social Research Council for funding the seminar series (Award Number R451 26 4775 98) from which this collection grew. It ran for two years from January 1999 in the Science and Technology Policy Unit at the University of the West of England and my thanks go to them for hosting the series and to Vicky Duggan, Jackie Nash and Christine Taylor for their enthusiasm and support. In particular I owe a huge debt of gratitude to my co-host, mentor and fellow researcher, Harry Rothman.

The peripatetic nature of the seminar series led us to the University of Central Lancashire, where thanks are due to Mairi Levitt of the Centre for Professional Ethics, and to the Science Policy Support Group at Birdcage Walk in London, with thanks to Peter Healey. One session was held with the Postgraduate Forum on Genetics and Society and I am most grateful for the organisational skills of Wan Ching Yee and Sandra Parsons. However the seminars would not have been a success without the commitment of those who agreed to give papers, as well as the many that came together in different combinations to form the audience.

I would like to thank Taylor and Francis, for permission to reprint a number of seminar papers that were subsequently published in the journal *New Genetics and Society* which Harry Rothman and I edit at Cardiff University. These are to be found in Chapters 2, 4, 5, 6, 7 and 13. I would also like to thank the publishers of *Public Understanding of Science*, IOP Publishing and The Science Museum, for permission to reprint the paper by Alan Irwin in Chapter 15.

Finally, I would like to thank Jackie Swift of the Cardiff University School of Social Sciences for her patience and professionalism in producing the final copy.

1 Introduction: What's New About the 'New Genetics'?

Peter Glasner and Harry Rothman

On 26th June 2000, President Clinton and Prime Minister Blair joined forces via satellite to announce the completion of the first draft of the map of the human genome to the expectant world. The project, begun in the mid 1980s in the teeth of much criticism from the scientific establishment (see inter alia Cook-Degan 1994) developed rapidly in the United States, Britain and some parts of Europe mainly through public and charitable funding. The goal, apart from satisfying scientific curiosity, was to increase the sum of human happiness through the amelioration of diseases caused by inherited gene disorders. As late as the beginning of 1998, most experts agreed that a complete sequence would not be available until 2005. However, the intervention of Craig Venter and the privately owned company, Celera Genomics, using a more selective and focussed approach backed by significant investment from venture capitalists, along with the threat to patent the resultant data, ensured that the timeframe was shortened significantly.

The result was supposed to be a blueprint of the three billion letter DNA alphabet (running to some three quarters of a million typed A4 pages). The reality is that about 90% has actually been sequenced as a rough draft, and of those discovered, only a handful relating to a few key single gene disorders have had their functions identified. This has led Nature Biotechnology (Editorial 2000) to suggest, in an editorial, that a new standard of human achievement has been set.

> At one stroke, musicologists have been able to declare Schubert's Unfinished Symphony "essentially complete", arguing that the great composer did write down all the notes he intended to use, albeit not in precisely the right order (and with some of the fiddly, boring bits left to be filled in by others).

Running in parallel with the Human Genome Project (HGP) has been the genetic modification of plants beginning with tobaccos, in 1983, and concentrating more recently on soybeans and maize. The goal of these interventions has been as laudatory as that for the HGP, namely increasing the sum of human happiness, but this time through efficiently feeding the World. Much recent research in the ethical, legal and social implications of the new genetics underplays the extent to which genomic information has become commodified knowledge in a globalised marketplace. This has a number of ramifications across both pharmaceuticals and agriceuticals.

To some extent American dominance is being challenged by European advances. By 2001 there were 2104 dedicated biotechnology companies in Europe compared to 1379 in the United States. However, it should be noted that the European firms tend to be smaller and poorer than their American counterparts – over half of European firms employ less than twenty staff (Nature Biotechnology 2002b). Despite this large scale entrepreneurial activity, it has to be remembered that biotechnology remains a speculative and high risk activity. Nature Biotechnology produces an annual business report of a portfolio of 440 public biotechnology companies, of which 20% are in Europe. Just 74 companies registered a profit in 2001, and the total loss for the portfolio was $5.3bn. Furthermore the most profitable companies were in the United States (Nature Biotechnology 2002a). In addition it is estimated that nearly 80% of global diversity is located in the global South but more than 80% of all patents granted in less developed countries belong to individuals or corporations located in the global North (Sagar et al. 2000).

The wider global issues are important, particularly when large multinationals such as Monsanto have been developing 'terminator' genes for key crops grown in countries like India, but also seeds resistant to all but their own herbicides, thereby forcing farmers into dependency on the manufacturers. It is also the case that some agriceutical companies are developing transgenic products to aid, for example, the reclamation of contaminated soils. It is possible that in the future a significant proportion of the world's needs for fuel, fibre, food and some medicines will develop from agricultural biotechnology. In this chapter, we will try to unravel the various ways in which these developing genetic technologies have come to be regarded as 'new'. Our discussion will begin by providing a brief overview of the science involved in an abbreviated historical context. This will be followed by an indication of how the new genetics developed as much of it now is, in joint academic and commercial environments, is becoming organised and managed in novel ways.

What is new genetics?

It may seem strange to ask this question, after all genetics is, arguably, a creation of only last century — barely a 100 years old as an academic discipline. Clearly it is not an ancient discipline, but neither is it, at a hundred years old, new. On the other hand advances in contemporary genetics are often headline news, so one may ask why is it that a science clearly well established, if not mature, makes the headlines more often than most other fields? In answering this question, or rather exploring it, we find ourselves forced to examine the interaction of scientific disciplines and their communities, the interaction between science and technology, the interaction between science and business, and lastly the interaction between science and the wider society. In fact we would argue as sociologists that genetics is an excellent locus for such explorations, perhaps more so than most disciplines. However, we hope to show in the course of these explorations, that contemporary genetics mutated and evolved sufficiently to make it sensible to talk of "the new genetics" if not "post-genetics". Here we are extending Fox-Keller's (2000) view that the gene concept is reaching the limit of its utility.

However, while this is not a historical chapter, some degree of historical perspective is necessary. Unfortunately given our constraints of necessary expertise and knowledge we might be in danger of presenting a Whig view of the development of genetics. We would advise readers, therefore, that our historical analysis is simplified and they should read specialist historians such as Robert Olby (1990) for a more nuanced picture.

It is convenient, though not totally accurate, to say that genetics has a single point of origin in the 1866 paper by Gregor Mendel; it was contemporaneous with Charles Darwin's great work "The Variation of Animals and Plants under Domestication" (1868).

Unfortunately, neither Darwin nor his scientific contemporaries understood the significance of Mendel's studies. Ernst Mayr (1982 p.682) comments that in the mid-nineteenth century "...the subject [of variation] was enveloped in great confusion. How difficult this subject is becomes apparent when one realises how bewildered even Darwin was...In retrospect it is obvious that much of it couldn't be clarified until after the rise of genetics (for example the distinction between genotype and phenotype)". The triple "rediscovery" of Mendel's laws in 1900 is possibly a more convenient starting date for genetics – the start of a new century. Clearly other biological work done in the late nineteenth century laid the basis for a rapid acceptance of Mendelism, for example, cytological studies of chromosomes, mitosis and meiosis etc.

Some writers have periodised the history of genetics, for example J.F. Crow (2000), and whilst such periodisations have a large degree of arbitrariness, they can help organise our thoughts providing we realise their limitations. We trust that in presenting such a periodisation scheme we have not forced the history of genetics onto a Procrustean bed of our ignorance and prejudices.

First period, 1900-43

During this period the genetics was dominated by breeding and cytological studies, when classical genetics made striking advances in transmission genetics. The term "genetics" was coined by William Bateson in 1905 and defined as "The elucidation of the phenomena of heredity and variation: in other words, to the physiology of Descent, with implied bearing on the theoretical problems of the evolutionist and systematist, and application to the practical problems of breeders, whether of animals or plants" (see Olby 1990 p.533). Olby notes that it was not just agricultural breeders who saw genetics as a fruitful scientific aid, "...there was a strong interest in the possible social applications of genetics. This eugenic concern was an important stimulus to the support of the subject in its early days, witness the Eugenics Laboratory at University College London (1906), the Balfour Chair of Genetics at Cambridge (1912) and the Eugenics Record Office at Cold Spring Harbour (1906)" (Olby 1990 p.533).

With hindsight we are able pick out key milestones in the history of classical genetics, by cutting through the historical undergrowth of debate and confusion which originally covered them. Suffice to say, within a decade or two, terms and concepts such as mutation, chromosomal inheritance, gene, phenotype and genotype, alleles, sex determination, linkage and maps, were established. New techniques, and experimental subjects, such as the use the fruit fly Drosophila, became part of the apparatus of genetics; by 1927 H.J. Muller was using radiation to induce mutations in fruit flies. As it matured classical genetics spawned sub-disciplines such as medical genetics, developmental genetics, and population genetics; the latter fulfilled Bateson's programme above by revolutionising evolutionary theory – what Julian Huxley termed "the modern synthesis" (Huxley 1942). Huxley's work contains a "progressive" eugenics programme, and his younger brother Aldous creates in "Brave New World" (1932) a pessimistic vision of a genetically engineered world – which continues to haunt. Yet the practical utility of genetics, part of Bateson's implicit manifesto was not achieved overnight. As late as 1939 H.C. McPhee is reported as stating at the Seventh International Genetical Congress, "Much has been written and spoken of the value of genetic research as the basis of livestock improvement, but little has been accomplished". However, by 1958, as Griliches (1958) demonstrated in his economic study of hybrid corn, measurable social returns were accruing from genetic research.

Second period, 1944-1953

In 1944 Oswald Avery and his co-workers discovered a process they called transformation, in which DNA transferred from one Pneumococcus bacterium could alter the hereditary characteristics of the recipient Pneumococcus cell. However most geneticists initially remained unconvinced that the molecular key to the gene was DNA, most were tipping proteins – far more complex in structure than nucleic acids.

We need to refer back to the early 1930s and the origins of molecular biology to understand why this was. Olby (1990, p. 518) warns against making the claim that there were two roots to molecular biology, the American "phage group" of Max Delbruch, and the British "structural school", and that the Watson-Crick model of DNA represented a synthesis of the two (Watson = phage group, and Crick = structural school). Olby prefers a broader conception of molecular biology drawing upon a number of research fields and methodologies. These included: three-dimensional structures of macromolecules; studies of the relation of such structures to biological function e.g. proteins; x-ray mutagenic studies; plant viral studies; and bacteriological and biochemical studies. There were also policy influences such as Warren Weaver's Rockefeller Foundation molecular biology research programme, conceived says Olby (p.505) as "...instrument dominated, interdisciplinary, and the most promising level of its analysis was ultra-structural." Clearly this approach was timely, C.P. Snow in his novel "The Search" (1934), describes Constantine's (aca J.D. Bernal) struggle to create an "Institute for Biophysical Research", adopting just such an approach, and even being advised to seek funding from "the Rockefeller people." "'(T)ell them this and this'... 'It's a pity it has to be so complicated' he grumbled ... 'I used to think human considerations didn't come into science...I admit human considerations come in, now...But how much easier it would be if they didn't'" (p.195).

By the early 1950s DNA began to be increasingly favoured as the key to understanding the gene. The story of how in 1953 James Watson and Francis Crick came to unveiled their double helix model of DNA, has reach mythic proportions – "the scientific discovery of the century", as well being the starting point for many takes on modern science, the triumph of opportunists, female discrimination and so forth. The "Double Helix" (Watson 1968) is nothing if not the story of "human considerations" – some might add and "inconsiderations" – in science. Be that as it may, it does mark the point at which an acceptable synthesis of genetics and molecular biology becomes inevitable. Thus Watson and Crick observe, in a masterpiece of understatement, in their landmark paper, "It has not escaped our notice that the specific pairing we have postulated immediately suggests a possible copying mechanism for the genetic material" (Watson and Crick 1953, p.737).

Third period, 1953-1973

James F. Crow (2000 p.810) says, with reference to the Watson-Crick DNA model, "Seldom in science does a single finding bring immediate order out of confusion". He notes that for a time the ensuing molecular biological research brought "...a great generality and simplification to genetics." He had in mind such discoveries as: messenger and transfer RNA, and the coding mechanism. However, it was not long before "...the complications started ..[although]. The deepest generalities remained... particulate inheritance in the early period, and the structure of DNA in the later..." "[B]ut questions of replication, transcription, translation, and regulation became more and more complex, and, still more distressingly, different from one species to another" (p.810). Consequently a mass of research programmes was spawned, together with new arrays of instrumentalities – sequencers, restriction enzymes, DNA libraries, clones, chromosome maps, PCR to name but a few. A surprisingly large proportion of these programmes and instrumentalities led to Nobel Prizes in Medicine and Chemistry, and sometimes they had implications beyond science and academe, opening the way for new technology, commodities and business profits: the rise of modern biotechnology. One technique, perhaps more than any other, might be taken as the source of modern biotechnology. This provided means whereby genes excised from one organism could be transferred to another. The procedures for doing this were developed in 1971-72 at Stanford University, various groups were involved, but for our story the key research involved the development of plasmid vectors (clones) by Herbert Boyer and Stanley Cohen in 1973. For, this signified not only a scientific advance, but also a technological and commercial turning point. Furthermore, it had profound implications for the organisation of genetic and molecular biology research, and for the relationship between science and the public. From their research came a landmark patent, US Patent 4,237,224 De. 1980 "Process for producing biologically functional chimeras", owned by Stanford University. However, as we'll see, the commercial significance of recombinant DNA was realised long before the Boyer-Cohen patent was public.

Fourth period, 1973-2000

The year 1973 is a convenient milestone date because it is the year that "human considerations" came into molecular biology with a vengeance! People began to be concerned that rDNA might create dangerous new organisms. Initially in 1974 this led to a voluntary moratorium by research scientists on some classes of rDNA experiments, and over the next few years, after much heated political debate, governmental regulatory research guidelines. These events have been analysed in Britain by Bennett *et al.* (1986) and the USA by Krimsky (1982).

These scientific activities began also to attract the attention of venture capitalists. The work of Boyer and Cohen was noticed by a venture capitalist Robert Swanson, who by 1976 had created, with Boyer, Genetech. While not the first new biotechnology start-up, Cetus preceded it in 1971, this created a new fashion for start-up companies, often spun off from academic centres, and involving many of the leading scientists in molecular genetics, and raising questions of scientific integrity. The science was changing in context as well as content.

It is a feature of scientific development for disciplines to split and to merge, and split again, often in a surprisingly unpredictable manner. For example, when genetics split from embryology at the beginning of last century, the fissioning of new disciplines, and cross fertilisation of disciplines, e.g. medical genetics in the 1930s, and the synthesis of classical genetics and molecular biology in the 1950s – which we have described above. But in this fourth period we find a coming together of molecular genetics and engineering in the form of genetic engineering or biotechnology. This is of more general theoretical significance because it may be considered as an example of a "technoscience", within which the traditional boundaries of scientific and technical knowledge production become blurred and contested.

The roots of biotechnology, however, go into prehistory when people discovered fermentation. Brewing played a major role in the 19th Century industrialisation process, and the science of fermentation and microbiology became by the late 19th Century important research fields. Further, as Robert Bud has shown (1993), even the term "biotechnology" has long and convoluted history, which long preceded its current incarnation. Such is the perceived economic potential of biotechnology that long-wave economists see it a key technology in a fifth Kondratief wave, which we are said to be currently moving into.

The 1980s saw great activity, political as well commercial, to grow biotechnology. National research programmes were created, ostensibly to enhance national industrial competitiveness. Hundreds of biotechnology start-up companies, mostly linked to medicine, were launched on the stock market. The returns were not always assured, for the investors it often proved risky and many firms bit the dust, were absorbed by the big 'pharma', or had to forgo any hopes of becoming independent pharmaceutical companies. Part of the problem was that to obtain investment funds these companies exaggerated the speed at which new products could be developed and reach the market. As with classical genetics, it took a decade or two for these practical applications to emerge. Classical genetics made, as we have seen, a contribution to agriculture, most notably the "Green Revolution" in the 1960s. The new agricultural biotechnology promised even more. This time the failure to deliver owed more to "human considerations" than technical difficulty. Genetically Modified Organisms (GMOs) became one of the most contested technologies of recent times.

In the late 1980s the Human Genome Project was created and promoted more for political and commercial reasons, than scientific (Cook-Degan 1994). It achieved its objectives of sequencing the human genome, at least in draft form. It had a simple target, but the achievement required big science investment, new organisational forms of research, and the creation of new powerful instrumentalities. It proved to be a novel mode of production for the life sciences. The official "completion" on June 26th 2000 was a great political occasion with prime ministers and presidents seeking reflected glory, and in the process inadvertently causing a crash in biotechnology stocks. With the HGP came new political and ethical concerns. This was foreseen by the organisers, who wishing to preempt the kinds of problems that dogged the early rDNA work, provided a small percentage of its funds for studying the ethical, legal and social implications (ELSI). Intellectual property rights again became a major issue, the contradiction of public and private found in the commodification of Nature in the early years of genetic engineering now took on a new cogency, when to nature we added to human. At this moment we have hundreds of pending patents for human sequences, with accusations of gene piracy and worse. These issues have split the scientific community.

Post-genetics, 2000 – the present

The completion of the HGP raises the question “where now?” The scientific discipline of genetics, which was born a hundred years ago, is surely a different beast today. It has matured in the way disciplines do, by mergers and splits, in the process growing quantitatively and qualitatively, creating new knowledge, and novel instrumentalities. It solved old problems by creating new problems. It extended its social reach by marrying technology and business, it has become a technoscience living off massive state public funding, and private investment, whose ownership is a tangled knot rivalling that of Gordius. Yet despite all the novelty and mutations its old agenda, adumbrated by Bateson, still remains, and the flawed eugenics agenda refuses to leave the stage, and veils itself in the pervasive notion of free individual choice.

Scientifically the new genetics may even, if some commentators are be believed, be in crisis as the limitations of the gene concept, and genetic reductionism become apparent (E. Fox-Keller, 2000). This battle is perhaps too early to call, a recent review of Fox-Keller in Nature says she is “...long on complaint and short on substance...” (Coyne 2000, p.26). Certainly the new genetics continues to demonstrate its disciplinary procreativity, currently spinning off, genomics, pharmacogenomics, proteomics, gene therapy and no doubt others. The HGP is over (almost) long live the next big programme! (Glasner 2002).

Organisation of the book

The book’s eighteen chapters are split into 5 sections, each with its own introduction and focusing on a different aspect of the new biosciences. The majority rely on a range of qualitative methods including ethnography, case studies, content analysis, discourse analysis, focus groups, interviews, and participant observation.

- Literacy, public understanding and the media: the Euroscreen-funded 'Gene Shop' – German attitudes to biotechnology – predictive medicine, new genetics and mental illness
- Commerce, industry and patenting: pharmaco-genetics and pharmaco-genomics – the right pill for the job – issues for public policy in commercialisation
- Gen-ethics: human genetic banking: banking of genetic material in Europe – regulation and social perceptions in Norway and Germany – proposed UK Population Biomedical Collection
- Genetic screening: implications for insurance – ethics and healthcare – choices and choosing in cancer genetics – genetic explanations in medicine
- Cloning, xenotransplantation and scientific democracy: Dolly the sheep – public attitudes to cloning in Britain – human and nonhuman identities; configuring scientific citizenship through public consultation

The collection will contribute to the development of qualitative methodology in an area where quantitative methods have proved less fruitful in exploring the contours and nuances of this developing new technology. Each chapter includes a bibliography and footnotes as well as, where necessary, a number of tables, figures and diagrams. There is a concluding chapter summarising the issues, and raising some of the new questions arising from the book which will need to be addressed in the coming decade.

References

Bennett. D., Glasner, P., and Travis, D. (1986) *The Politics of Uncertainty. Regulating recombinant DNA research in Britain*, London: Routledge and Kegan Paul.

Bud, R. (1993) *The Uses of Life. A history of biotechnology*, Cambridge: Cambridge University Press.

Cook-Degan, R. (1994) *The Gene Wars: Science, Politics and the Human Genome,* New York: Norton.

Coyne, J. A. (2000) Review of 'The Century of the Gene' *Nature*, 408, p. 26.

Crow, J. F. (2000) Epilogue. Genetics yesterday, today and tomorrow. A personal view, in D. P. Snusted and M. J. Simmonds *Principles of Genetics*, New York: Wiley.

Editorial (2000) First draft of genome sets new industry standards, *Nature Biotechnology*, 18, p. 803.

Fox-Keller, E. (2000) *The Century of the Gene*, Cambridge MA: Harvard University Press.

Glasner, P. (2002) Beyond the Genome: Reconstituting the New Genetics, *New Genetics and Society*, 21, 3, pp. 267-277.

Griliches, Z. (1958) Research costs and social returns: hybrid corn and related innovations *Journal of Political Economy*, 16, p. 419.

Huxley, J. S. (1942) *Evolution, the Modern Synthesis,* London: Methuen.

Krimsky, S. (1982) *Genetic Alchemy. The social history of the recombinant DNA controversy*, Cambridge MA: MIT Press.

Mayr, E. (1980) *The Growth of Biological Thought*, Cambridge MA: Harvard University Press.

Nature Biotechnology (2000a) Public biotechnology 2001 – the numbers, *Nature Biotechnology*, June, pp. 551-555.

Nature Biotechnology (2000b) Supplement: Bioentrepreneurs, *Nature Biotechnology*, July, p. BE3.

Olby, R. C. (1990) The emergence of genetics, in R. C. Olby et al. (eds) *Companion to the History of Modern Science*, London: Routledge.

Sagar, A., Daemmrich, A., and Ahiya, M. (2000) The tragedy of the commoners: biotechnology and its publics, *Nature Biotechnology*, 18, pp. 2-4.

Watson, J. D. (1968) *The Double Helix*, London: Wiedenfeld and Nicholson.

Watson, J. D. and Crick, F. H. C. (1953) A structure for deoxyribose nucleic acid *Nature* 171, pp. 737-8.

PART I
LITERACY, PUBLIC UNDERSTANDING AND THE MEDIA

2 The Gene Shop at Manchester Airport

Mairi Levitt

Public understanding is promoted in European countries as a necessary condition for informed debate on the uses of genetic technology. The Gene Shop aimed to promote public understanding and debate by providing reliable up to date information and a staff of specialist health visitors and doctors who could answer questions and enter discussions with their knowledge not only of the science of genetics and current applications in health care, but also the practical, social and ethical problems that arise for families and individuals with a genetic disorder. Currently, most people find out about genetic disorders and testing only during pregnancy or after the birth of a child with a genetic disorder into their family. The Gene Shop endeavoured to attract people who were not in one of those crisis situations. It formed part of a project investigating ethical, legal and social implications of genetic screening and testing, which focused on the areas of commercialization of testing, insurance and public education.

The Gene Shop was in Terminal 2 departure lounge between the Body Shop and the Prayer Room. The terminal was used for charter flights rather than business travel and therefore had roughly equal proportions of men and women and a low proportion of middle class passengers (one-fifth were social class A and B according to the airport's own figures, 1996). To encourage people to approach them, staff wore badges with their first names on and specially designed Gene Shop sweatshirts. Owing to the logistics of staffing, the shop was open for 6 hours a day on 5 days a week, whereas the airport was normally open 24 hours. In the year it was open, 10,500 people visited.

A poster in the Gene Shop written by Dr Maurice Super, a clinical geneticist, summarized its aims:

- Easily accessible public education
- Reduction of mystique and media hype
- To satisfy and encourage curiosity about genetics and the surrounding issues
- Together with the Centre for Professional Ethics to evaluate the role and value of such a Shop

X No counselling as such.

The prior assumptions of the Gene Shop were: that the public lack information on genetics and genetic testing; that they are, or could be encouraged to be, curious to know more; and that their views might be affected by 'media hype'. These are assumptions common to public understanding of many science initiatives. Recent initiatives refer to "inaccurate media coverage", ignorance of what genes are and the difficulties of informing people and the need for an education programme aimed at producing "genetic literacy" (Richards, 1996; Walpole *et al.* 1997).

The content of the Gene Shop was decided by medical staff and included four interactive touchscreen programmes previously used in the Wellcome Centre's 'Genes Are Us' exhibition and the Science Museum's Genetic Choices exhibition. Three of these presented case studies of people with a genetic disorder, for example, sickle cell anaemia and cystic fibrosis. The presentations were generally optimistic about the future, and the message was that advances in genetics would solve the problems soon, for example through gene therapy for cystic fibrosis. The eight permanent poster displays had details of the science of DNA/genes, modes of inheritance, a Royal Family tree and real-life examples and photographs of sufferers from particular disorders with a diagram of the relevant genes. A small room off the main room was used as an office and to show videos, particularly for group visits. At the back of the shop there was an illuminated chromosome display and a section with materials on topics, which changed monthly. Here charities associated with different disorders presented a positive image of affected individuals with photographs and case studies. The contrast between information provided in the Shop on detecting disorders such as Down's prenatally, with the option of termination since there is no treatment, and that on specific people with Downs provided by the Downs Syndrome Association, illustrated Lippman and Wilfond's discussion of "twice-told tales". She refers to two different stories, one told to parents deciding whether to abort a foetus affected by a genetic disorder and another to parents who have had a baby with the same disorder (Lippman and Wilfond, 1992). The contrast between the different stories was not referred to.

The Gene Shop was funded by the European Commission as a demonstration activity "undertaken to transfer results from technology producers to technology users". "Demonstration projects must have a strong user's need orientation and will always involve the participation of the medical profession"; it had to be evaluated "under realistic operation conditions" (European Commission, 1994, pp.12, 52-3).

Methodology

As far as possible, the evaluation had to be carried out without interfering with the activity of the shop, which was designed to attract passers-by and allow people to browse and only speak to staff if they wanted to and in complete anonymity. Staff recorded the number of visitors, their age group and sex, and logged queries and comments with age group, sex and, if people volunteered it, their occupation. In addition, questionnaires for adults were left in the shop and offered by staff. For a more systematic evaluation, every adult visiting on five selected days was asked to complete a questionnaire or interviewed using the same questions while trying not to discourage visitors (as one teacher put it "if people have got to answer questions from people like you they might not come in, it would be better if they were just free to browse" – which on most days they were). A total of 263 adult questionnaires were completed and 21 'passers-by', who were in the departure lounge but had not come into the shop, were also interviewed to see if they differed in terms of background, knowledge and interest. The 15 members of staff completed questionnaires anonymously. A children's questionnaire was completed by 225 children age 3-16 and those findings are discussed in the full report (Levitt, 1998). The evaluation was limited by reliance on short interviews and questionnaires and the inability to do a follow-up study to monitor any long-term effects, since more intensive interviews or requests for further contact would have compromised confidentiality and the casual open access policy of the Shop.

Evaluation findings

The Gene Shop was evaluated in terms of its aims.

Aim 1: Easily accessible public education

The Shop was certainly accessible in the sense of being free, situated in a place where people had time to spare and open to anyone who wanted to visit, including non-travellers such as school parties and families who came to the airport for a day out by train or car. Asked about the advantages and disadvantages of the Gene Shop as a source of information, the most frequent comments among both visitors and staff were that it was accessible to the casual visitor, there were knowledgeable and/or friendly people to question and the information was of high quality, particularly the interactive computes and other visual materials. Visitors also commented positively on the airport location as a place where people have time to spare and "health on their mind". Those who cited disadvantages, 14% of the visitor sample, thought the Gene Shop would be more accessible in a different location or as a mobile unit or that the depth of information might confuse other people and could cause anxiety.

A crucial part of the evaluation of this project was to find out about the characteristics of people visiting the Gene Shop and their motives for coming in. The Gene Shop was in a good position to attract those other than the professional middle class, only 21% of passengers were social class A or B, and there were equal numbers of males and females because of the lack of business travellers who are over 80% male at Manchester airport (statistics provided by the airport). Despite this, those who actually came into the shop on the five interview days were disproportionately professional middle class (44%), nearly one-fifth were students and only 5% had manual occupations. This would not have been avoided if the Gene Shop had been situated in a shopping centre with a similarly mixed population. In contrast, the social class profile of the sample of passers-by was closer to that of the passengers as a whole, with the largest group, 38% in manual occupations.

The number of visitors on any one day ranged from 134 down to two, but there was no clear correlation between the numbers and the time people spent in the Shop. The highest number of visitors came when a busy flight to Florida was delayed. Most adult visitors were in the child-bearing years for whom genetic information is particularly relevant, and around a fifth of visitors were children. Just over half the adult visitors were female but, among the children, there were more boys than girls; boys seemed particularly attracted by the touchscreens. The time people spent in the Shop was monitored on the five 'interview' days. The average times ranged from 11.5 minutes to 4 minutes for women and from 7.6 to 5.5 minutes for men. Children were usually with an adult and so spent similar lengths of time in the Shop. Groups from schools, hospitals and colleges on prearranged visits would stay for longer.

Information on why people came to the Gene Shop came both from a direct question for those who filled in a questionnaire or were interviewed and from observation by the Gene Shop coordinator who worked full time in the shop. By these methods, adults with different kinds of interest were identified, and these have been given as six groups. Numbers based on the survey of visitors give precise figures but those based on staff impressions are necessarily imprecise.

Professional interest in health A large proportion of general enquiries about the shop came from individuals who had a professional interest in the subject matter, including doctors, nurses, teachers, scientists, students, patient support group members and hospital managers. Almost a third of the visitor sample had jobs with a 'professional interest' in health. Such visitors showed a good deal of interest and enthusiasm and felt that their fellow professions/students could learn from the Shop. The group had some knowledge of science and genetics already and could follow up an interest or talk to the staff. They are a section of the general public who are likely to be attracted to educational activities and tend to share the scientific and medical perspective. The interest of this group, together with that of the staff who reported learning from the displays, shows the potential for information targeted at those working in health and education.

Specific enquiries There were over 300 enquiries about 95 different diseases and conditions, many were related to a personal family history; one third were multi-factorial. Here the staff could give specific information and one in 65 instances individuals were given details about their regional genetic service and how they could be referred to it. The genetic diseases which were most commonly asked about were cystic fibrosis and cancer, probably influenced by the material available in the shop in the monthly displays.

Previous experience A smaller group of visitors had some previous direct experience of a genetic condition and/or genetic counselling. Typically they had had experience of genetic disease in the past but were no longer involved in the genetic services because their child had died and/or they were no longer of childbearing age. For them the Gene Shop offered an anonymous, 'safe' place to talk about their experience to knowledgeable staff, rather than to seek any further information. Staff commented that this group were often experts in the condition that affected their family, having had to work hard to acquire the practical knowledge they needed to deal with the situation.

This raises the question of what knowledge these people would have wanted to be easily available to them at the relevant time. In research with parents of children with Down's, Layton *et al.* Contrast the motivation for seeking knowledge of scientists and the parents (Layton *et al.* 1993). Whereas scientists are curious about the natural world and have the long-term goal of understanding, parents wanted knowledge that was relevant to what needed to be done for their children. The knowledge parents were offered reflected the different priorities of the scientists. Thus scientific knowledge of the causes was not relevant and of little practical use to new parents.

Anxious/concerned There were a small number of visitors who felt concerned that things were moving too quickly in the field of genetics, for them the Gene Shop provided a place to air their worries and fears.

> What is it all leading to? Will we as a society end up selecting out people who have certain diseases? Health resources are not a bottomless pit but the more we find out the more we have to spend. Don't want people messing about with genes.
>
> (adult, male)

Although individuals expressed opposition to aspects of genetic technology, they did not show hostility to the Shop itself. It may be that a Gene Shop in a High Street would have attracted the opposition of local organisations, e.g. opposed to abortion. In the airport, most people visited as individuals or families and the presence of staff who people could talk to, materials showing positive images of genetic disorders (e.g. from self-help groups), and the emphasis on choice may have defused any hostility as intended.

Fascinated by genetics A very small group of people, described by the coordinator as 'memorable', found the whole subject area of genetics fascinating and were often well read. For them anything and everything on genetics was of interest, and they tended to have lengthy discussions with staff.

General interest The final group was both the largest and the least well served by the Gene Shop and similar public education initiatives. These visitors seemed to have a general interest in the shop and its contents and were usually travellers waiting to fly. In interviews and questionnaires most respondents reported that they had come "just to fill in time" or to see what a Gene Shop is. Most did not speak to the educators but those who did tended to be relating the material to themselves rather than expressing an interest in the science. The shop was designed to attract this group and colourful displays gave the initial impression of being accessible, but those who looked more closely found difficulties. For example, a map of Europe showed the incidence of one mutation of cystic fibrosis, which would tend to be of interest to geneticists, rather than the incidence of carriers of any mutation.

Aim 2: Reduction of mystique and media hype

In the deficit model of public understanding the public are found to be ignorant of genetics as measured by questions devised by scientists, and to have irrational fears about its applications fuelled by sensationalising media coverage. There was little evidence that visitors had fears about genetics brought about by media coverage. Among all the visitors who spoke to staff as well as those in the sample, all except a few showed interest rather than concern over the latest developments in genetics that occurred when the Shop was open and received extensive media coverage, including Dolly the sheep and headless tadpoles. Since media coverage of new developments in genetics relies heavily on the scientists' own accounts the idea of media hype should not be accepted unquestioningly. The sources of information most often mentioned by the visitors in the sample were television and/or radio but often with the added comment "I watch it if it is on". Around half the visitors had spoken to their family and friends about genetic conditions or disease and a quarter had spoken to a GP.

The implication of concern over media hype is that accurate knowledge of the science will lead to a more positive view of its applications. However, among visitors and passers-by, there was evidence of uncritical support for genetics and its applications rather than irrational fears. It was the staff whose work involved day-to-day contact with families affected by genetic disorders who were the least enthusiastic. All were asked for their opinion and comments on eight statements, four of which were positive and optimistic about the impact of genetic technology on health, and two were more cautious, although they were not asked in this order.

1. I want to know as much as possible about any genetic disease I might develop.
2. I would want my children tested to see if they will develop a genetic disease.

3. I want to know about testing for my partner and me, to see if our children might be born with a genetic disease.
4. We will be healthier when the genes for more diseases are found and tests are developed.
5. I would only want to be tested for genetic diseases which can be treated.
6. You can get so many things wrong with you. Who wants to know?

The last statement was a comment made by a woman whose close relatives had tested positively for a breast cancer gene (Brown, 1993, p.37).

There was very little difference in responses between visitors and passers-by (the first two columns in Figure 2.1). Generally, both had a positive attitude towards genetic testing even when the diseases might not be treatable, while the staff were less positive. Younger people, age under 25, were the most positive and the small group of those age 60-plus were the least positive. The desire for knowledge among the majority may reflect a desire to know even bad news about their health, a confidence that untreatable diseases will become treatable or an optimism that they will not themselves have an untreatable genetic disease. The figure of 57% who wanted to know even if the disease was untreatable contrasts with the small percentage of individuals who are at risk of Huntington's Disease who have chosen to be tested (Marteau and Richards, 1996, p.5).

On child testing for diseases that develop later, those age 25-45 who were most likely to actually have dependent children, had the lowest level of agreement but still a majority in favour of testing (57.6%). Within families, mothers would probably be organising such testing for their children, and mothers had slightly higher levels of agreement than fathers (69%: 60%). However, only one member of staff would want child testing for his own children.

The general optimism was shown by answers to the statement "we will be healthier when the genes for more diseases are found and tests developed". Tests on their own will not make us healthier but most respondents put 'yes' to the statement. Even of those with a professional interest, three-quarters agreed with the statement. They were perhaps expressing enthusiasm for the research and/or a conviction that tests would soon lead to treatment. This view reflects early media coverage of new tests, which tends to include optimistic statements from the researchers involved (Chadwick and Levitt, 1997).

Over half the respondents added comments in the 'I think' speech bubble provided and 62% were positive, 22% expressed worries about genetic research or testing, and the remainder were neither positive nor negative, e.g. mentioning a family history. Among the negative comments were general worries, often expressed after a positive comment, and three specific concerns: abortion, child testing and the possibility of genetic knowledge leading to discrimination. The themes found in the positive comments were the importance of genetic knowledge in general, including praise for the Gene Shop, and the value of genetic testing in particular. A typical comment, from a female school teacher was "The more information available the more informed choices there are".

Figure 2.1 Opinions on genetic research and genetic testing

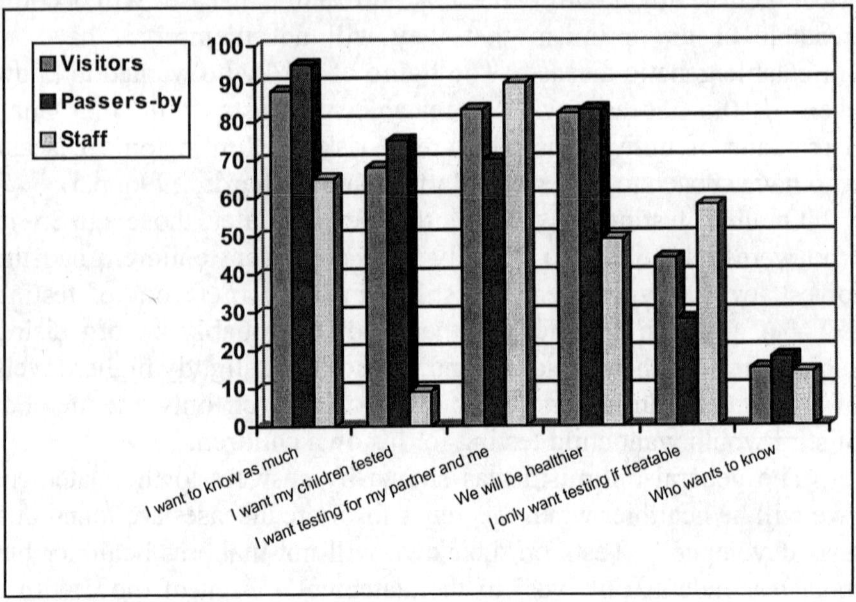

Aim 3: To satisfy and encourage curiosity about genetics and the surrounding issues

The visitors to the Gene Shop were asked to assess their own knowledge and interest in genetics, genetic disease and health on a scale of 1-5. Generally, interest was rated higher than knowledge and the greatest level of interest was claimed for health. There was no class difference in 'interest in health', but those in middle-class occupations were more confident of their knowledge. Not surprisingly, those with any form of professional interest in genetics and genetic disease, including doctors, nurses and health visitors, gave high ratings, with men consistently higher than women for all except 'interest in health'. Those in the oldest age group (65-plus) rated themselves lowest for knowledge of genes and of genetic disease. Interest in genetic disease was highest in the under-25s and the 25-45 age groups, who would be most likely to have young children or be planning a family. Passers-by rated themselves lower than visitors on everything except interest in health. These figures do at least show that people want to present themselves as very interested in health, which would include genetic disease if it affects them personally (directly or indirectly).

X No counselling as such

The Gene Shop was set up as a public information facility, which was to be non-directive and not represent any sectional interests. The staff had been to briefing sessions, which emphasized their role as educators rather than counsellors and that was how they described their role in the shop. Five of them felt they were also promoting genetic services. However, the Gene Shop might have been perceived by visitors as trying to promote a particular view of genes and genetic testing. The sample of visitors were therefore given a series of statements and asked which "applies to the Gene Shop in your opinion?"

Figure 2.2 Self-assessed knowledge and interest (rated 1 low to 5 high). Sample of visitors N=256

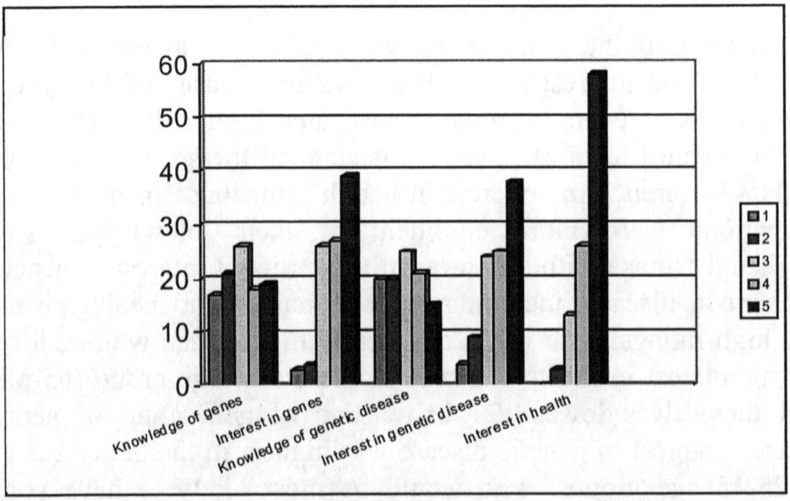

The most neutral statement was the most popular: "the Gene Shop is intended to give information on genes and genetic tests" (98% agreed). Most people also agreed that "the Gene Shop is designed to interest people in genes and genetic disease" (96% agreed). Most rejected a more active view of the Gene Shop's purpose and did not agree that "the Gene Shop is intended to encourage people to have genetic tests" (63% disagreed). Visitors agreed that "the Gene Shop does not represent any particular point of view" (74% agreed) and did not think that "the Gene Shop only represents the point of view of the medical profession" (79% disagreed). Those interviewed tended to back up their disagreement by citing the information from families experiencing genetic disease, which was provided by the touchscreens and displays.

The Gene Shop agenda

Public education initiatives including those by the Wellcome Trust and Euroscreen provide information on the science of genetics, genetic screening and gene therapy in an attractive and positive format (e.g. leaflet on planning a baby "Play the odds in your favour"). The rhetoric is that public education provides neutral information so that the public can make informed choices. However, information is necessarily selective, and interactive touchscreens control both the information and the responses. Assessment of attempts to consult the public such as the National Consensus Conference on plant biotechnology, have shown how the presentation of material structured the framing of questions by the lay public (Durant, 1994; Barns, 1996). Genetic research and testing raise problems that are not addressed by more information on genetic technology – one which Anne Kerr examined with focus groups was "where to draw the line with genetic research and testing". Research with adults and children has shown that a lack of technical knowledge does not prevent sophisticated discussion of the social and ethical implications of the new genetics (Kerr, Cunningham-Burley and Amos, 1998; Levitt, 1999). However, the "drawing the line" exercise could not be conducted with the public health professionals who formed one group. They were "unusually antagonistic" to the first stage in which participants discussed their views on genetic testing and research generally and then looked at vignettes, e.g. of a couple facing choices about cystic fibrosis testing. Many of their responses were "defensive or dismissive" (Kerr *et al.* 1998, p.123). In other word they were uncomfortable, and therefore hostile to engaging in debate. At the same time, the sorts of fears raised by lay people and the media are often dismissed by health professionals as irrational and media hype (Kerr *et al.* 1997). If these fears/problems are taken seriously (rather than dismissed as "never going to happen" or something that was done only in the past-Nazi eugenics) then harder simply to dismiss them. A similar attitude was found among staff at the Gene Shop: when asked what they thought people needed to know more about, they chose the science of genetics but not social and ethical implications.

The Gene Shop displays tended to invoke enthusiasm and admiration for genetics and the medical profession, who enjoy high ratings of trust in the UK, but it reinforced the boundaries between lay people and experts by concentrating on areas where the professionals felt safe.

References

Barns, I. (1996) Manufacturing consensus: reflections on the UK National Consensus Conference on Plant Biotechnology, *Science as Culture*, 5, 2 (23), pp. 199-216.

Brown, P. (1993) Breast cancer: a lethal inheritance, *New Scientist*, 18 September, pp. 34-8.

Chadwick, R. and Levitt, M. (1997) in R. Chadwick, M. Levitt and D. Shickle (eds) Mass Media and Public Discussion in Bioethics, *The Right to Know and the Right not to Know*, Avebury: Aldershot.

Durant, J. (1994) *Final Report. UK National Consensus Conference on Plant Biotechnology*, Science Museum: London.

European Commission (1994) Biomedicine and Health Research (Biomed 2) 1994-1998 Workprogramme, European Commission: Science Research Development.

Kerr, A. Cunningham-Burley, S. and Amos, A. (1997) Drawing the line: an analysis of lay people's discussions about the new genetics, *Public Understanding of Science*, 7, pp. 113-133.

Kerr, A. Cunningham-Burley, S. and Amos, A. (1998) The new genetics: professionals' discursive boundaries, *Sociological Review*, 45 (2), pp. 279-303.

Layton, D., Jenkins, E., MacGill, S. and Davey, A. (1993) *Inarticulate Science: Perspectives on the Public Understanding of Science and some Implications for Science Education*, East Yorkshire: Studies in Education Ltd.

Levitt, M. (1998) *The Gene Shop. Evaluation of a Public Education Facility*, University of Central Lancashire.

Levitt, M. (1999) Drawing limits: contemporary views on biotechnology, *Journal of Beliefs and Values*, 20 (1), pp. 41-50.

Lippman, A. and Wilfond, B.S. (1992) Twice told tales: stories about genetic disorders, *American Journal of Human Genetics*, 51, pp. 936-7.

Marteau, T. and Richards, M. (eds) (1996) *The Troubled Helix*, Cambridge: Cambridge University Press.

Richards, M. (1996) Lay and professional knowledge of genetics and inheritance, *Public Understanding of Science*, 5, pp.1-14.

Walpole, I. R., Watson, C., Moore, D., Goldblatt, J. and Bower, C. (1997) Evaluation of a project to enhance knowledge of hereditary diseases and management, *Journal of Medical Genetics*, 34, pp. 831-7.

3 Public Understanding of Genetic Engineering in Germany

Jürgen Hampel

Introduction

At the turn of the century, biotechnology is one of the most controversial topics in Europe. Human cloning, research with human embryo stem cells, genetically modified food, all these applications of biotechnology raise serious public controversies (Grabner *et al.* 2002). But controversial debates in Europe about biotechnology are not only a recent development. They have a long history (Torgersen, Hampel *et al.* 2002). And within Europe, Germany is one of the forerunner countries of critical views on genetic engineering. In Germany, very fierce debates on biotechnology could be observed from early 1980s up to the present, with changing main focuses, from contained use of genetically modified organisms to gm-food and cloning. Numerous NGOs organised the protest against applications of biotechnology and up to the late 1990s, the destruction of experimental fields with genetically modified plants was common. Not only the intensity of the public and political debates was very high from the beginning, the German public is also said to have particularly negative attitudes towards biotechnology. On the other hand, Germany succeeded to be one of the most innovative countries in relation to biotechnology. If the number of start-up companies is considered to be the criterion of assessment, Germany has become the leading biotech-country in Europe in the late 1990s.

The question is, how these divergent developments, the long history of critical and fierce debates and a sceptical public on the one hand and the economic development of biotechnology on the other hand fit together. To answer this question, public opinion in Germany in relation to biotechnology and the German debates on biotechnology are analysed. In a series of research projects, both on the national and international level attitudes towards biotechnology, media reporting about biotechnology and societal and political controversies and regulatory outcomes have been analysed. A large scale German project with nine different sub-projects has made in depth analyses of public opinion, published opinion and their interrelationship (Hampel and Renn, 2000). Collaboration in the International Research Group on Biotechnology and the Public allowed to extend the scope of research also to the international level (Durant, Bauer and Gaskell, 1998; Gaskell and Bauer, 2001; Bauer and Gaskell, 2002), where they were allowed to analyse whether observed developments are specific for Germany or whether they have a common European or international background.

Before public perception of biotechnology will be analysed, as a first step, the various debates on biotechnology are discussed in the context of the historical and social development of technological controversies in Germany to provide an understanding of the cultural and political backgrounds of the development of public opinion and public debates in Germany.

German debates on technology and biotechnology

Being one of the most advanced countries in the development and application of technology, Germany has also a long tradition of critical thinking about the future consequences of technological development, from the Romantic Period up to the present (see Huber, 1991). The historically unique situation in the first decades following World War II led to an immense support of new technologies in Germany. In that period, technological development has been seen as a major cause of the so called "Wirtschaftwunder", the rapid growth of the post-war-economy in Germany. Up to the 1970s, new technological developments had been legitimised by a common understanding that new technology means progress. Technology has been seen as a guarantor of better living conditions and a better quality of life. But the overwhelming acceptance of technology has disappeared. As can be shown by longitudinal studies, the 1970s and 1980s experienced a considerable decline of public acceptance of technology (see

Kliment, Renn and Hampel, 1994; Renn and Zwick, 1997). This change did not lead to an increased rejection of technology, but to balanced attitudes where both benefits and losses of technical development are reflected.

One of the most important and the most fierce controversies in the seventies could be experienced in conjunction with nuclear energy, where the anti-nuclear movement succeeded to stop the further implementation of nuclear energy in Germany. The anti-nuclear movement succeeded also in creating new political and scientific institutions which are shaping debates on technology up to now. The Green Party which started as an environmentalist party has become a steady and meanwhile widely accepted member of the German party system. At the same time, the Öko-Institut in Darmstadt and Freiburg, a new scientific institute opposed to traditional science has been founded to make environmental research independent of government and industry and to support the resistance towards nuclear energy with scientific knowledge. Also other technological developments raised controversial debates in Germany. In the early eighties, Germany experienced a fundamental debate on new communication and information technologies, which culminated in the controversy about the German National Census in 1987.

The beginning of the German debates on biotechnology

While biotechnology was only of little public interest in the 1970s and early 1980s, two triggers started an intense debate in Germany; the birth of the first German test tube baby and the attempt of Hoechst to establish a plant for the production of human insulin using methods of genetic engineering (Hampel *et al.* 1998). The way the German public reacted to the first out-of-laboratorium applications of biotechnology was similar to reactions to the implementation of other new technologies. This conflict seemed also to be very fundamental and of a greater fierceness than in other countries.

Although biotechnology became a public topic only in the first half of the 1980s (Torgersen, Hampel *et al.* 2002), legal regulation of biotechnology has started earlier. Already in 1978, when biotechnology was of interest only for experts in that field, first regulations of biotechnology which were closely related to the American NIH-guidelines from 1976 had been implemented. But in that period, biotechnology remained an academic topic with almost no public resonance (Hampel *et al.* 2001; Torgersen, Hampel *et al.* 2002).

While first regulations relied on professional self-control, biotechnology became a political topic in the early 1980s, when biotechnology changed from being an academic research tool to become an applied science with applications in agriculture, medicine and pharmaceutics. Applications which raised serious debates (Gill, 1991). In that period, the Green Party entered the German Parliament for the first time.

For the discussion of the problems related with modern reproduction medicine and genetic engineering, two commissions had been founded in 1984, the Benda Commission, discussing the problems related to reproductive medicine and medical applications of genetic engineering and the Parliamentary Enquete Commission on "Chances and Risks of Genetic Engineering" to discuss non-medical applications of genetic engineering. While medical biotechnology was discussed both in a risk and an ethics frame, the regulatory discourse on agricultural biotechnology was restricted to a risk frame, where only arguments based on scientific concepts of risks were allowed. The recommendations of both commissions led to two laws, the German Gene Law (Gentechnikgesetz) and the Law for the protection of Human Embryos (Embryonenschutzgesetz), both had been passed in 1990. While the German Gene Law was closely related to the two European guidelines 90/210/EWG and 90/220/EWG, but more restrictive, the Law for the Protection of Human Embryos, which strictly prohibited any consuming use of human embryos or human embryo stem cells, had no European counterpart. Both regulations differed in their ability to close public debates. Different to the Law for the Protection of Human Embryos, which was accepted as a binding regulation fulfilling the regulatory needs for almost one decade, the German Gene law was criticized from both opponents and supporters of biotechnology. While industry and science evaluated the Gene Law as being too restrictive and to endanger the scientific and economic development of biotechnology, opponents argued within the risk frame – the only frame for critical invoices which was allowed by the regulation – but attempted to change the basic risk concept from a traditional scientific understanding of risk to new concepts of risk (Beck, 1986; Krohn' Krücken, 1993; Bonβ, 1995) which address uncertainties in an unknown future.

The economic paradigm – biotechnology as basic innovation

After the German unification with its enormous economic costs, economy as a topic gained importance in public and political debates. Biotechnology was discussed as the basic technology for the economic development of the next decades. As a consequence, the economic perspectives of biotechnology became a prominent frame for the discussion of biotechnology (Torgersen, Hampel *et al.* 2002). Supporters of biotechnology argued that lagging behind international developments would endanger the economic perspectives of the country.

In the 1980s and early 1990s, Germany had a backlog in commercial applications of biotechnology, in spite of having a strong scientific basis in molecular biology. A first attempt to commercialise biotechnology in Germany, the attempt of the company Hoechst to establish a production plant for human insulin, failed because of legal reservations (Thielemann, 1998). But in general, the German chemical and pharmaceutical industries failed to recognise biotechnology as an innovative business sector, although there had been hints from the administrative system as well as from science that genetic engineering and biotechnology would offer new fields for innovation (e.g. Dolata, 1995). So the major impulses for development in biotechnology came from the German Ministry for Research and Technology (BMFT) and were funded with public money.

As a consequence of the increasing importance of the issue of economic competitiveness of Germany after the German reunification, the German Gene Law, which was criticized as being too strict and therefore to endanger the competitiveness of the German Industry, was reformulated in 1993 using any opportunity given by the European prescriptions to make the German law less restrictive (Hampel *et al.* 1998).

To reach the goal of the German government to become the No.1 in Europe in the commercialisation of biotechnology, the German Ministry for Research and Technology (BMBF) started a very successful program in 1996, the Bio-Regio-contest. This program invited regions to develop concepts in order to improve the social, political, technical, economical and administrative situation of new start-up companies in their regions and to form and to support innovation networks. This contest fostered intense activities in Germany which have been observed by the media and commented in a positive way. Critical reactions from the public could not be observed. The ongoing attempts of the German government to establish a biotech industry in Germany finally were successful when in the second half of the 1990s Germany experienced a remarkable biotech-boom, mainly in the field of medical and pharmaceutical biotechnology. The market turnover of pharmaceuticals produced using methods of genetic engineering doubled from DM 882.9m in 1996 to DM 1,848.5m in 1999 (Informations-Sekretariat Biotechnologie, www.i-s-b.org). The whole German biotech-industry has doubled its turnover in the year 2001, according to the result of a study of the Landesbank Baden-Württemberg. According to the Ernst and Young Report 1999, Germany had the largest number of start-ups in the life-science-industry. The general number of companies was only surpassed by the UK (Ernst and Young, 1999). But this report demonstrates also that the average size of the companies in Germany is quite small.

GM-food and Dolly – triggers for new debates

The observed biotech-boom describes the development of new companies in the area of pharmaceutical and medical biotechnology. But the increasing economic importance of biotechnology did not result in a general decrease of social and political debates about biotechnology.

When the first ships with genetically modified soy beans reached Germany in November, the 6th 1996, Greenpeace activists started to campaign against genetically modified food. Food biotechnology increasingly became a topic of public concern, but raised no heated debate. Intense media reporting could not be observed before January 1997 (Hampel *et al.* 2001). Although the pressure on regulators increased and led to the Novel Food Prescription Act, which has been passed by the European Commission in January 1997, the societal activities were of higher importance for the further development than the political and regulatory activities, which remained in the frame of scientific risk analysis and did not meet the growing need for public participation (Dreyer and Gill, 2000; Hampel *et al.* 2001). Greenpeace organised the public resistance and forced supermarket chains not to market gm-products. In consequence, instead of the fact, that there was no official withdrawal of genetically modified food from the German market, almost no products labelled as genetically modified are available. The attempt of a food company to sell a chocolate bar in Germany being labelled to contain genetically modified ingredients directly evading supermarkets failed and was stopped in July 1999. But although the willingness of the German population to consume genetically modified food is very low, the topic is not very salient (Hampel *et al.* 2001). It seems that industry and political regulators follow a duck and cover strategy not to provoke public reactions.

Different to the development of agricultural biotechnology was the situation in the field of medical applications of biotechnology, where the Law for the Protection of Embryos which has been passed in 1990, met the regulatory needs of the public which saw no need for further debates.

This situation did not change substantially, when the birth of Dolly the sheep, the first mammal cloned with adult cells, was announced. When in most European countries, as a consequence of the Dolly-debate, the problem of the regulation of the use of human sperm cells and human cloning initiated an intense public debate (Einsiedel *et al.* 2002), the reactions in Germany were rather modest (Hampel *et al.* 2001). The expected cloning of humans appeared on the agenda in Germany, but it was more a debate in the arts sections of the newspapers than a political debate: in sum, the impact of this debate seems to be rather low. In April 1997, a commission of the Ministry for Research and Technology directed by the head of the German Research Association (DFG) came together, including representatives from science, law, humanities and theology. It evaluated the existing regulation in Germany, the German law for the protection of embryos, as a sufficient regulation of the problems which had been raised by the technical opportunity of cloning. So, different to politicians in other countries, German politicians could refer to the existing regulation in Germany (Hampel *et al.*, 2001).

The existing regulation, which calmed down the public debate, raised on the other hand the resistance of scientific and medical institutions, which considered this regulation as being too strict and hindering research necessary for therapeutic purposes. From time to time, individual researchers as well as research associations lanced the demand of weakening of the strict German regulation and adapting the German law to the international standards to enable German scientists and start-up companies to be competitive with the international scientific community. A debate on the change of this regulation started. A starting point was a gap in the law for the protection of human embryos, which prohibited the production of human embryos or human stem cells for any other purpose than human reproduction, but did not consider the import of human embryonic stem cells. The new debates were focussed not only on risk issues and ethical issues, but also on economic perspectives and scientific progress.

While these debates did not evoke the interest of the general public, a lecture of the German philosopher Peter Sloterdijk in July 1999 dealing with Martin Heidegger and his letter on humanism (where Heidegger claimed that humanism has not reached its goal to civilise man), raised a serious debate which resulted in a series of articles in the weekly ZEIT and in other elite newspapers. TV-channels changed their programs and broadcasted discussions on the opportunities of modern biotechnology in the human field. In his lecture, Sloterdijk mentioned "Regeln für den Menschenpark" (rules for the human park) and talked about "anthropotechnologies". He asked the question, whether mankind will end in changing its mode of reproduction from "birth fatalism" to optional birth and prenatal selection. Critics accused Sloterdijk of supporting eugenic applications of biotechnology and the breeding of humans with applied biotechnology. The topic of human reproduction was on the agenda and remained on the agenda of the public and political discussion.

To support the political decision making processes of both parliament and government, new institutions had been founded to discuss the problems related to biotechnology. To enforce the scientific discussion of the ethical basics of biotechnology, the German Reference Center for the Ethics of Bio-Sciences has been founded in Bonn. To support the political discussion, the German Parliament established a Parliamentary Enquete-Commission "Recht und Ethik der modernen Medizin" (Law and Ethics of Modern Medicine). In parallel to the parliamentary Enquete-Commission, the German Chancellor established another committee, the National Ethics Council (Nationaler Ethikrat) to discuss the problems of modern bio-medicine. Although the recommendations from these institutions were rather unclear, the German Parliament allowed the import of human embryo stem cells in 2002, when they are created before January 1st 2002. Again, this decision has closed the debates on biotechnology even if not all the actors involved are satisfied with it.

Public perception of biotechnology

Germany is said to be one of the most special countries in relation to biotechnology. And indeed the history of biotechnology in Germany – the failure of gm-food in Germany, the long lasting controversy, the destruction of experimental fields – indicate that the German public holds negative attitudes towards biotechnology. But on the other hand, there is a remarkable biotech-boom, which makes

Germany one of the leading biotech-countries with the highest number of start-up companies in Europe. Different to the situation in the 1980s, the establishment of biotech-facilities do not raise public concerns.

Looking at public perception of biotechnology, a first glance seem to support the view of a particularly sceptical German public. Eurobarameter 1996 shows as well as the Eurobarometer surveys from 1991 and 1993, that support of biotechnology was lower in Germany than in most other European countries (Gaskell *et al.* 1998; Hampel *et al.* 1998). But at second glance, the image of a German public resistant to biotechnology cannot be supported. Indeed, it can be shown, that Germans have less positive expectations about biotechnology than the average European, and than the populations of most other European countries. Compared with the 45 percent of Europeans who thought in 1996, that genetic engineering or biotechnology will positively influence their lives within the next twenty years, the rate in Germany (36.2 percent) is rather low. But from the opposite perspective, in its expectation that biotechnology will make life worse, the German public was even in 1996 not very different from the average European. Admittedly, the proportion of Germans thinking that genetic engineering will have negative effects on their lives (23.2 percent) was higher than the European average (19.2 percent), but people in other northern and central European states were more concerned about the further implications of biotechnology than were the Germans. With 18.3 percent higher than in any other member state of the European Union was the share of people thinking that genetic engineering will have no effects for their own lives. Another fifth (22.3 percent) were not able to express an attitude towards genetic engineering.

That attitudes of the German public towards biotechnology are more or less ambivalent was also a result of the German Biotech-Survey from 1997 (Hampel, Pfenning and Peters, 2000), where the questions were differentiated according to level of approval and disapproval. The picture obtained was depicting indecision and uncertainty. One fifth of the persons interviewed saw themselves unable to make an assessment of genetic engineering. Of all persons interviewed who have already formed an opinion on genetic engineering, about 40 percent assessed genetic engineering ambivalently, i.e. as equally good and bad. About 20 percent respectively tended to reject genetic engineering and approve of it with some reservations, only 6-7 percent turn out to be strict opponents or supporters. The ambivalent assessment of genetic engineering in the mid 1990s was also shown by the fact that 44 percent of the persons interviewed in the 1997 survey assume that chances and risks are balanced.

Germans were not driven by fears but by scepticism about the opportunities offered by applications of biotechnology. Looking at the evaluation of the likelihood that several positive and negative developments will occur – for example, that genetic engineering will create new diseases or that genetic engineering will help to solve world hunger problems – it can be shown using the data of the Eurobarometer 1996, that not only regarding positive expectations, but also regarding negative expectations, the Germans ranked below the European average (Hampel *et al.* 1998).

That the evaluation of biotechnology was not fundamentalist in nature is also demonstrated by the fact that only a few people used the extreme points of the attitude scales and by the fact that Germans evaluated the importance of biotechnology lower than did most Europeans. On a 10-point scale (10 is extremely important), we got an average of 6.1 in Germany, while the European average was 6.5. Even the proportion of people thinking that modern biotechnology is extremely important was lower in Germany (6 percent) than in most of the other European countries.

The unique situation of Germany, which showed, together with some other countries like Austria, the least support for biotechnology in Europe up to the mid 1990s, changed, when in the second half of the 1990s, after the first import of genetically modified soy-beans in 1996, a heated debate on agricultural biotechnology swept over Europe (Grabner *et al.* 2001). This debate even increased its intensity, when a few months later, in February 1997, the birth of the first cloned mammal, Dolly the sheep, was announced and biotechnology became one of the most dominant topics in public discourses in Europe (Einsiedel *et al.* 2002).

While other European countries, especially in Southern Europe, changed their support of biotechnology and became more sceptical, partly with dramatic shifts (Gaskell and Bauer, 2001), the German public remained quite stable in this period (Hampel *et al.* 2001). Germany shifted from being one of the most sceptical countries in 1996 to the European average in 1999 (Gaskell and Bauer, 2001).

In Germany, as in 1996, biotechnology and genetic engineering are, compared to other technologies, also in 1999 the technologies with the most critical reactions of the respondents. In 1999, the general attitudes of the German population towards biotechnology and genetic engineering are very similar to 1996: 37.9 percent express positive expectations, 19.7 percent, a slightly smaller share than in 1996, express fears about the future consequences of the development of genetic engineering and biotechnology. One fifth of the respondent, still a higher share than in any other European country, thinks that biotechnology and genetic engineering will have no effects on their own lives and 22.4 percent of the respondents are unable to estimate the consequences of genetic engineering.

Although biotechnology became more and more a part of everyday life in the years following the 1996 Eurobarometer, it is surprising, that almost one third of the respondents declare that they haven't talked about biotechnology before the interview. Although developments of biotechnology raised severe controversies and intense media reporting, the share of people not having communicated about biotechnology increased from 24.1 percent in 1996 to 31.4 percent in 1999. That biotechnology is not a main topic for everyday communication in the public is also supported by the non-awareness of the various concrete applications. Looking at different applications, the share of respondents who never heard about these applications is surprisingly high, from 32.2 percent, who never heard about the use of modern biotechnology in the production of foods, for example to make them higher in protein, keep longer or change to taste up to 68 percent, who never heard anything about genetically modified bacteria to clean up slicks of oil or dangerous chemicals.

Unlike genetic engineering in general, where we can find high uncertainty and somehow mixed feelings, different applications are evaluated more strictly. As in older surveys, also in the Eurobarometer survey from 1999, different applications of genetic engineering are evaluated in very different ways. While medical and pharmaceutical applications find strong support in the German public, most people (up to 80 percent) reject applications in agriculture, especially in food production and animal breeding. The relationship between general attitudes towards genetic engineering and attitudes towards specific applications of genetic engineering is rather low (Hampel and Pfenning, 1999). General views on genetic engineering are far from determining views on specific applications.

Perception and evaluation of 'red' genetic engineering

Although the medical application of Genetic Engineering was one of the starting points of the German debate (Gill, 1991), the support of medical applications of biotechnology is very high. In the 1999 Eurobarometer survey, almost 70 percent (67.2 percent) of the respondents think, that the introduction of human genes into bacteria to produce medicines or vaccines, for example to produce insulin for diabetics, should be supported, a slight increase compared to 1996. Genetic Testing to detect diseases we might have inherited from our parents are supported by 69 percent of the respondents (64.2 percent, 1996). Even therapeutical cloning, which was not part of the 1996 questionnaire, find support in the German public (50 percent support). Compared to 1996, the support has become even stronger.

But it would be misleading to concede that medical applications are supported in general. Supportive attitudes become less prevalent, if genetic engineering deals with animals, even in the field of medical biotechnology. In the 1996 Eurobarometer, the genetic modification of animals for xenotransplantations was less supported than any other application including genetically modified food. In 1999, the cloning of animals such as sheep to get milk which can be used to make medicines and vaccines is supported by only 35 percent of the respondents. More than 50 percent (50.3 percent) of the German respondents reject this application. As with Xenotransplantations in 1996, the cloning of animals gets less support than any other application of biotechnology. The Eurobarometer survey from 1999 allows us to look in detail at the evaluation of the cloning of animals. The evaluation of this application combines both lack of benefit and a high perceived catastrophic potential. Forty-one percent of the respondents strongly support the statement, that this application of genetic engineering is simply not necessary. The catastrophic potential of this application is seen as very high. About 40 percent claim, that it threatens the natural order, a statement which is only rejected by 10 percent of the respondents, 38 percent strongly agree to the statement, that it would be a catastrophe, if anything went wrong and 37 percent have a strong feeling of fear when thinking of this application.

Perception and evaluation of 'green' genetic engineering

While most 'red' applications of genetic engineering are supported by the public, the so called green biotechnology raises substantial resistance in Germany, and not only in Germany, but also in Europe (Gaskell *et al.* 2001).

While taking genes from plant species and transferring them into crop plants to make them more resistant to insect pests is supported by almost 50 percent of the respondents, the application of modern biotechnology in food production is supported by only 36.4 percent. The strong decrease of food biotechnology, which could be found between 1996 and 1997 in Germany (Hampel, Pfenning and Peters, 2000), cannot be found comparing the results of the 1996 with the 1999 survey in Germany. But in other European countries, particularly in southern Europe, the support for agricultural biotechnology decreased substantially (Gaskell and Bauer, 2001).

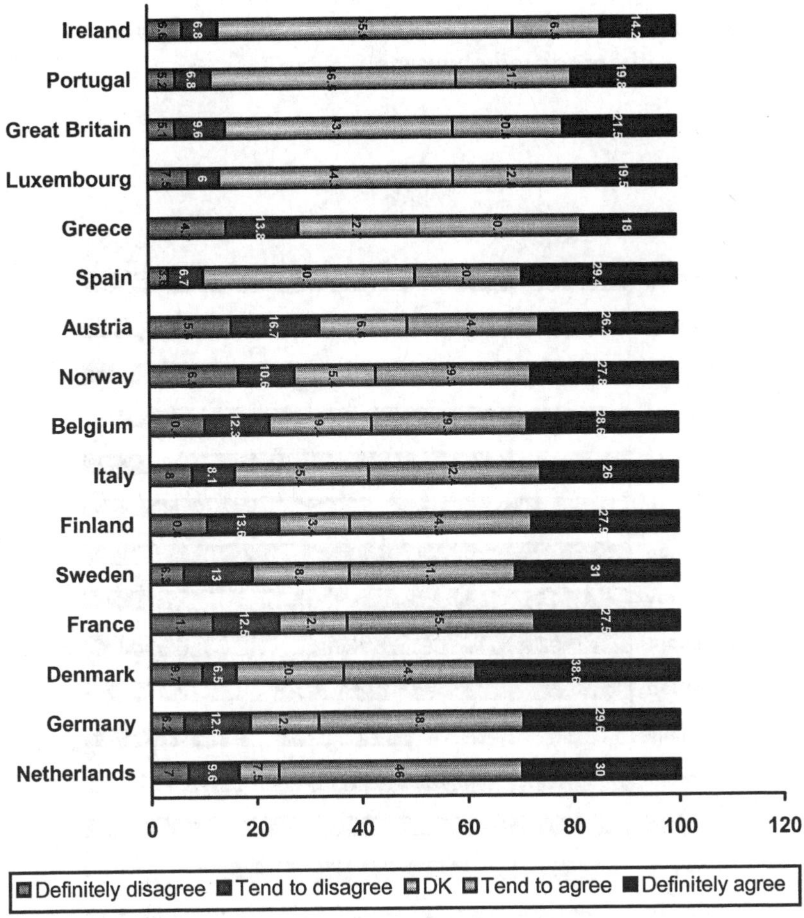

Figure 3.1 Evaluation of biotechnology – Use of biotechnology to produce medicines should be encouraged

Considerably lower than the support of these applications is the personal willingness to consume genetically modified products, which can only be found with about 20 to 25 percent of the German respondents. There is almost no differentiation between GM products containing genetically modified organisms like GM fruits or whether the final product does not contain any genetically modified substances like GM sugar. But as with medical biotechnology, support for agricultural biotechnology becomes even lower when animals are involved. Only 20 percent would tend to consume eggs from genetically modified chickens.

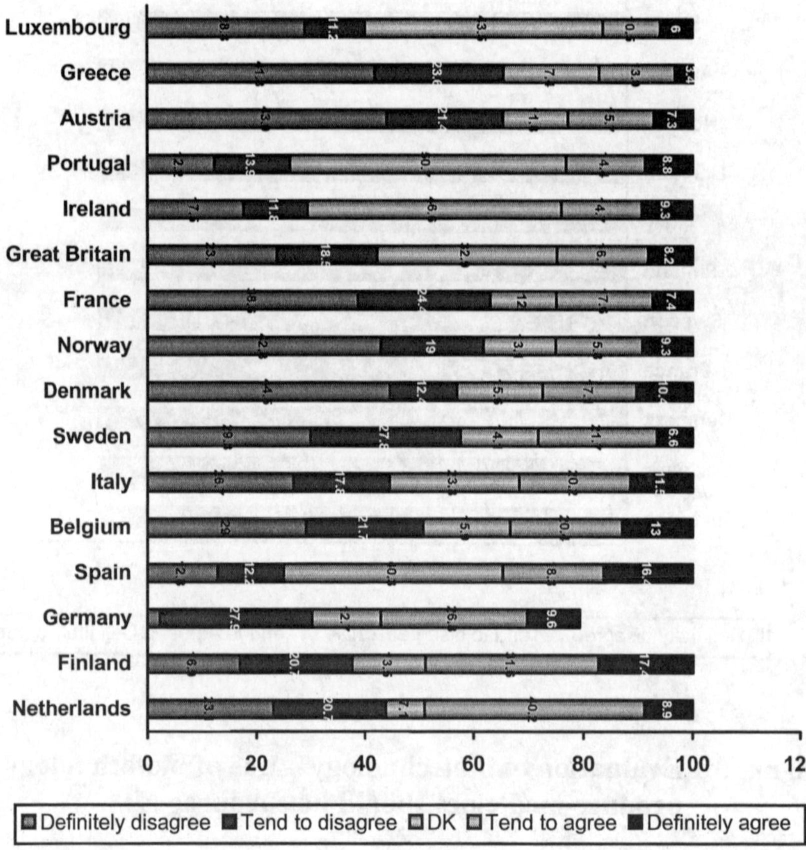

Figure 3.2 Evaluation of biotechnology – Use of biotechnology to produce food should be encouraged

Only a small minority, 8.2 percent thinks that genetically modified food will bring benefits to a lot of people. 18 percent slightly agr ee, but almost 50 percent of the German respondents disagree with this statement. Almost 70 percent of the respondents (67.8) fear that a catastrophe would appear if anything went wrong. More than 60 percent of the respondents dread the idea of gm food and 59 percent (58.6 percent) reject that GM food is no danger to future generations. Only 5 percent of the respondents think that they are on the safe side with biotechnology. And a remarkable 15 percent strongly reject, that they can avoid the risks associated with GM food.

For a strong minority GM food is not a question of democratic decision making neither of more sensible and gradual introduction. 47 percent of the respondents reject that GM food should be allowed, when a majority of the public would be in favour of it and 35 percent do not agree that GM food should be introduced more gradually. These answers indicate, that for a strong minority, GM food is not a question of judgement but of more general considerations.

Cognitive backgrounds of evaluation of biotechnology

Amongst natural scientists involved in public debates about genetic engineering, it is a common explanation that negative attitudes towards genetic engineering in general or applications of genetic engineering are a result of a lack of knowledge. If people would have sufficient and correct information, so is the assumption, there would be a breaking down of 'irrational fears'. Indeed, people in Germany know very little about genetic engineering. At least they think so. In Germany, according to the Biotech-Survey, only 1.5 percent think that they are very well informed about genetic engineering. 12 percent of the German respondents of the Eurobarometer 1999 think that they are sufficiently informed about biotechnology. The lion's share thinks they have only moderate or little knowledge. But even if knowledge is low, we did not find in older surveys like the Eurobarometer 1996 and the German Biotech Survey a systematic relation between the individual knowledge and attitudes on genetic engineering. Different to these earlier surveys, we find in the 1999 Eurobarometer survey effects on education and knowledge both on the support of food biotechnology and on the willingness to consume genetically modified food. But this does not mean, that there is a direct cognitive relation between knowledge and support.

Lack of knowledge can also be found by supporters of genetic engineering. This lack of knowledge is not only stated on the individual level, but also and this is more important for the discussion, on the scientific level, a stated lack of knowledge, especially knowledge of long-term effects and side-effects, of the experts themselves (Peters, 1999). The state of information is not only described by the term 'we don't know', but also by the term 'they don't know'. The available expert knowledge is not seen as being sufficient for a secure evaluation of genetic engineering. The application of genetic engineering is seen as an experiment where the outcomes cannot be foreseen.

This stated uncertainty about the expected and unexpected outcomes of genetic engineering is leading to the next question, the risk evaluation of genetic engineering. Genetic Engineering in general is seen as being risky. Large parts of the public thinks that it has a high catastrophic potential (s.a.). But surprisingly, the evaluation of risks and benefits is more or less balanced. And in the general acceptance of risks, we do not find substantial differences between supporters and opponents of genetic engineering.

In the 1999 Eurobarometer, we asked for six applications whether they are useful, risky, morally acceptable and whether they should be supported. We find very high correlations between the evaluation of usefulness, moral acceptability and support of the respective application of biotechnology. The correlation coefficients between these three indicators are in any case with the exception of food biotechnology higher than 0.7, which means, that more than half the variance of these different criteria is common. The correlation between the risk evaluation and the other criteria is substantially lower, with the exception of the applications in agricultural biotechnology lower as 0.4. In the case of food biotechnology the relationship between the evaluation of the usefulness and the support of this technology is lower than 0.7 (0.68). Looking at this application, the risk dimension is of higher importance for the support than with the other applications (-0.44). But when we control for the other criteria, the perceived benefits and the ethical judgement are dominant, not the perceived risk.

This suprising result can be explained by a deeper analyses of the data. All analysed surveys, the Eurobarometers as well as the German Biotech-Survey, find nevertheless substantial differences between supporters and opponents, which explain the lack of statistical influence of the risk dimension in the regression model. While opponents share unanimously the opinion that the risks of genetic engineering are outweighing the potential benefits, supporters are living in greater cognitive tensions. They are divided into two groups of the same size, into optimists thinking that the benefits outweigh the risks and in a precarious group thinking that the risks and benefits are balanced.

If we ask about the risks of genetic engineering, scientific or social risks are not dominant. Concrete risks do not seem to be that important, dominant are unspecific risks, also the uncertainty about the type of risks which might be associated with genetic engineering and also references to the German history during the twelve years of fascist rule. In general, the dominant risk of genetic engineering is its novelty, people think being involved in an experiment where nobody knows the outcome. The risk dimensions quoted by the public do not fit to traditional risk concepts, for example the technical risk concept. New concepts of risk, for example by authors like Ulrich Beck (1986), Wolfgang Krohn and Gerhard Krücken (1993) and Wolfgang Bonβ (1995) are more appropriate to understand what people think when they talk about risks of genetic engineering.

As already stated, we can observe enormous differences in the attitudes towards specific applications of genetic engineering. These differences can also be stated when we look at the benefits people expect with the use of genetic engineering. Here we can observe a strong division between medical and pharmaceutical applications on the one hand and agricultural applications on the other hand. In the medical field, people expect that genetic engineering will contribute to the development of therapies of diseases like Cancer and Aids. If these expectations will come true, it is worth accepting some risks. The situation is totally different when we look at applications in agriculture. If people refer to 'benefits' of genetic engineering in this field, they usually remark the absence of benefits. The goals guiding the application of genetic engineering in agricultural genetic engineering are not supported by the German public. At the moment, people perceive only internal benefits of the use of genetic engineering in agriculture, an increase of the productivity of agricultural production, not benefits for the consumer. And the increase of the productivity of agricultural production is not seen as being beneficial but more as being a threat. In food production, 'natural' is seen as a sign of quality. This evaluation does not only reflect different views on food, which should be as natural as possible, but also a trust in food production, in scientists and in regulators.

This refers to another general problem of genetic engineering, the lack of trust. Trust in scientific experts is very low, only 3% of the respondents of the German Biotech-Survey think, that one can have full trust to scientific experts because of their knowledge (Hampel, Pfenning and Peters, 2000). Also, the independence of scientists is questioned by a large majority. Rather similar is the share of respondents who think that Universities can be trusted most in questions of biotechnology. They are not seen as being independent and they are not seen as being able to make an overall assessment of the consequences of biotechnology. We can also see, that social and political institutions are seen as being rather powerless with regard to the development and application of biotechnology. Dominant actors are science and industry (Hampel, Pfenning and Peters, 2000).

And the legal regulation of genetic engineering is not seen as being sufficient. It is very important to state that the lack of trust in the social embedding of genetic engineering can not only be observed when looking at opponents of genetic engineering, but also people with positive attitudes are very concerned about the regulation and control of genetic engineering. Here, we can observe an enormous trust gap.

It is commonplace to assume that there is a strong media effect on the attitudes of the Public towards Genetic Engineering. And in Germany, technical innovators as well as political advocates of the technological development process complain that the German 'Sonderweg', is due to a critical reporting of journalists in the

German press. But research on media reporting indicates that this common view is not true (Kohring *et al.* 2000).

Biotechnology is not only a point for media reporting, it is also discussed in private networks. In the German Biotech Survey, where not only analysed was how people estimate the public opinion, but also the opinion climate of their social network. An interesting result is that people estimate the public opinion more sceptically than public really when we are looking at the individual data. So supporters of biotechnology see themselves more as a deviating minority than the supporters of biotechnology. But, in contradiction to theories like the 'Schweigespirale' (spiral of silence) from E. Noelle-Neumann, this estimation of deviating from the General Public has no effect on the willingness to communicate about biotechnology. Both supporters and opponents see themselves as being concordant with their own social network (Hampel, Pfenning and Peters, 2000).

Outlook

We started the analysis with the question, how the critical public climate in Germany regarding biotechnology and the German biotech boom fit together. Empirical analyses demonstrate that these developments, which are at first glance contradictory, fit together. We find a field of generally accepted applications and we find substantial economic developments in that field. On the other field, the economic importance of agricultural biotechnology, the rejected field of application of biotechnology is low. So economic importance and public perception are not contradictory, they fit together.

Empirical analyses also demonstrate that common explanation for the reluctance of the German public cannot be supported by data. There is not an overwhelming and emotional resistance but a weighing of perceived benefits, ethical consideration and perceived risks. As a result, some applications are supported, others not. Only a minute minority assesses all applications of genetic engineering uniformly. As a rule, the same person will both approve of some applications and reject others.

Above all, applications in the field of medicine are supported, whereas applications in agriculture are rejected by a large majority. This result was already determined in previous studies (Kliment *et al.* 1995). But even within these fields of application there are some considerable differences. Not all applications of genetic engineering in agriculture are rejected and not all medical applications of genetic engineering are approved of. Thus the question for the reasons of approval and rejection arises. Owing to the qualitative in-depth investigations, which asked for associations and reasons, we can let the persons interviewed speak for themselves.

Genetic engineering above all raises apprehensions in the context of humans, fears that humans could be manipulated in their basic dispositions and qualities and be bred. Environmental risks, which dominate the 'public' discussion, come up as topics much more rarely. But it is precisely the applications on humans which are accepted, whereas applications on plants and, much more so, applications of genetic engineering on animals meet with rejection. This is a contradiction only at a first glance, as the applications on humans, which give rise to the greatest fears, namely manipulation of the human sperm cells or the cloning of humans, are decidedly rejected.

Genetic engineering is not only seen as a high-risk technology, it is also raising hopes. Hopes of decisive breakthroughs in the therapy of diseases such as cancer and AIDS, hopes of new drugs. Where these hopes exist, where there is a positive utopia of genetic engineering, we also find great approval, such as in therapeutic applications.

Where there is no such positive utopia, genetic engineering is regarded indifferently to critically. The frequently emphasized contribution of genetic engineering to the solution of the problem of global hunger is recognized in part, even if it is considered a problem by many. Even if the application of genetic engineering is approved of for this purpose, it is far from the point of general acceptance of genetic engineering applications in agriculture in Germany. To date promoters of genetic engineering have failed to produce an accepted reason for such an application of genetic engineering. Here, genetic engineering does not only appear unnecessary, but even as a threat to product quality. Hence it does not surprise that, above all those agricultural applications of genetic engineering aiming at the goal of 'an increase in productivity' are decidedly rejected.

Little also speaks for the idea that the acceptance of one application also leads to the acceptance of other applications. Even if applications within one field of application are more closely linked amongst one another than with applications from other fields, the correlations between the assessments of concrete applications among each other as well as between the assessments of concrete applications and the assessment of genetic engineering as such are weak at best.

On the application level it is the expected benefit and its assessment, which more than the perception of risks evokes acceptance or non-acceptance.

Acceptance, if present, is cognitively not as stable as rejection. Proponents of genetic engineering experience greater cognitive tension than opponents of genetic engineering. Whereas the latter are convinced that genetic engineering is a high-risk biotechnology think, that there are not only benefits, but also risks and that risks and benefits are somewhat balanced.

Here it must be considered that the risks of genetic engineering as seen by the public can only be covered inadequately with the classical technical term of risk. It is not technical risks that are in the foreground, but social risks on the one hand, the fear of an abuse of genetic engineering, on the other hand insecurity about future consequences. The non-arrival of damaging events in the past is only to a certain extent suited to weaken the assessment that risks do exist. It must be emphasized, however, that the perception of risks does not have a compelling influence on their assessment. Risks are absolutely accepted when they are put up with in exchange for something which appears to be worth it. If this is not the case, there is no occasion to put up with even the slightest of risks.

Proponents of genetic engineering also see themselves more strongly in contradiction to the climate of public opinion, whereas opponents of genetic engineering are rather of the opinion that they are in agreement with the majority of society. But these assessments do not have an effect on the personal communication behaviour, because people seeing themselves in discordance with the general public, experience that their social network supports their view.

Both on the individual and the social level knowledge about genetic engineering is considered inadequate. Only a few of the persons interviewed in the Biotech Survey consider their knowledge to be very good. A two-thirds majority is of the opinion that their knowledge of genetic engineering is rather deficient. Little knowledge was also determined in those projects which dealt in-depth with individual aspects of the assessment of genetic engineering. There is only little knowledge about molecular biology in general and genetic engineering in particular. And the existing knowledge is not always correct.

The question however is, how the knowledge about genetic engineering and the assessment of genetic engineering are linked. Behind many 'educational campaigns' one can find the implicit conjecture that reservations with regard to genetic engineering can be attributed to deficient knowledge. Based on this assumption, there would be no rejection of genetic engineering if everyone had the knowledge of genetic engineers. One must take leave of this idea. It has already been pointed out that people generally have little knowledge about genetic engineering. This, however, is equally true for supporters and opponents of genetic engineering. Proponents do estimate their knowledge slightly higher than critics, but the correlation between knowledge and attitude is a very weak one. However, the smaller the knowledge, the greater is the probability that the persons interviewed do not have an opinion.

The evaluation of Genetic Engineering cannot be reduced to technological or scientific aspects. It is a process where technological developments and their applications are proved whether they are compatible with dominant value in a society. For this reason, the conflicts on genetic engineering in Germany are primarily conflicts, where, in the sense of a reflexive modernization, concepts of modernization are evaluated by the public.

As can be shown by looking to the argumentation patterns in the controversy on genetic-engineering, this debate is not a technical debate but a debate focussing on social, economical, ethical and ecological subjects. Along with new technological opportunities the basic relation between man and nature is as questioned as the relation between the scientific and political elites and the population (Hampel and Renn, 1999). Under this perspective resistance towards a new technology can be seen as resistance against an undesired future development. So discussions on new technologies are reflecting the problems modern society have with the coordination and regulation of the different subsystems (Bauer, 1995).

The development of the public debates on biotechnology in Europe, which started in some countries and swept over to most European countries after 1996, and the situation in 1999, where Germany has become an average country in Europe regarding perception of biotechnology indicates, that Germany, as well as other forerunner-countries of debates on biotechnology, is not a country with a genuine 'German' way of perception of biotechnology, but a country, like others, especially in Northern and Central Europe, where people discussed earlier than in other countries the expected benefits, but also the expected risks of biotechnology.

The future of biotechnology in Europe is open. We find public support in medical biotechnology, but the most important debates in the last years were on medical biotechnology. So it is not unlikely that people become aware of the opportunity, that their expectation, that biotechnology may contribute to the development of new therapies against cancer and other diseases may be accompanied by the perception of developments which challenge human identity or the fundaments of modern societies, from selection of embryos to the differentiations in the opportunities of assurance.

On the other hand, the rejected applications of agricultural biotechnology may get increased support if there is a clear regulation for labelling giving individuals the opportunity to decide if benefits are directed to the consumers and not the producers of food and if the ability to deal with ecological and health risks is increasing.

The future of biotechnology will depend not only on the further development of this technology, but even more on the ability of societies to provide structures which take up and deal with public concerns.

References

Bauer, M. (1995) 'Towards a functional analysis of resistance' in M. Bauer (ed.) *Resistance to new technology – nuclear power, information technology and biotechnology*. Cambridge: Cambridge University Press, pp. 393-417.

Bauer, M. and Gaskell, G. (2002) (eds) *Biotechnology – the making of a global controversy*. Cambridge: Cambridge University Press.

Beck, U. (1986) *Vom Risiko. Unsicherheit und Ungewißheit in der Moderne*. Hamburg: Hamburger Edition.

Dolata, U. (1995) 'Nachholende Modernisierung und internationales Innovationsmanagement – Strategien der deutschen Chemie – und Pharmakonzerne' in Th. Von Schell and H. Mohr (eds) *Biotechnologie-Genetechnik. Eine Chance für neue Industrien*. Berlin, Heidelberg: Spring, pp. 456-480.

Dreyer, M. and Gill, B. (2000) 'Germany: 'élite precaution' alongside continued public opposition', *Journal of Risk Research* 3 (3), pp. 219-226.

Durant, J., Bauer, M. and Gaskell, G. (1998) (eds) *Biotechnology in the the Public Sphere. A European Sourcebook*. London: Science Museum.

Einsiedel, E. et al. (2000), 'Brave new sheep – the clone named Dolly' in M. Bauer and G. Gaskell (eds) *Biotechnology – the making of a global controversy*. Cambridge: Cambridge University Press.

Ernst and Young (1999) *Communicating Value, European Life Sciences 99, Sixth Annual Report*. London: Ernst and Young.

Evers, A. and Nowotny, H. (1987) *Über den Umgang mit Unsicherheit. Die Entdeckung der Gestaltbarkeit von Gesellschaft*. Frankfurt/Main: Suhrkamp.

Gaskell, G. and Bauer, M. (2001) (eds) *Biotechnology 1996-2000. The years of controversy*. London: Science Museum.

Grabner, P. et al. (2001) 'Biopolitical diversity: the challenge of multilevel policy-making' in G. Gaskell and M. Bauer (eds) *Biotechnology 1996-2000. The years of controversy*. London: Science Museum, pp. 15-34.

Hampel, J. et al. (1998) 'Germany' in J. Durant, M. Bauer and G. Gaskell (eds) *Biotechnology in the Public Sphere. A European Sourcebook*. London: Science Museum, pp. 63-76.

Hampel, J. et al. (2001) 'Biotechnology boom and market failure: two sides of the German coin' in G. Gaskell and M. Bauer (eds) *Biotechnology 1996-2000: The years of controversy*. London: Science Museum, pp. 191-203.

Hampel, J., Pfenning, U. and Peters, H.P. (2000) 'Attitudes towards Genetic Engineering' in German Attitudes to Genetic Engineering, special edition of *New Genetics and Society*, Vol. 19, No. 3, pp. 233-250.

Hampel, J. and Renn, O. (2000) (eds) German Attitudes to Genetic Engineering. Special edition of *New Genetics and Society*, Vol. 19, No. 3.

Huber, J. (1991) *Technikbilder*. Opladen: Westdeutscher Verlag.

Kliment, T., Renn, O. and Hampel, J. (1995) 'Die Chancen und Risiken der Gentechnik aus der Sicht der Bevölkerung' in Th. von Schell and H. Mohr (eds) *Biotechnologie-Gentechnik. Eine Chance für neue Industrien.* Berlin: Heidelberg, Springer, pp. 558-583.

Kohring, M. and Görke, A. (2000) 'Genetic Engineering in the international media: an analysis of opinion leading magazines' in German Attitudes to Genetic Engineering: Special Edition of *New Genetics and Society*, Vol. 19, No. 3, pp. 345-364.

Krohn, W. and Krücken, G. (1993) 'Risiko als Konstruktion und Wirklichkeit. Eine Einführung in die sozialwissenschaftliche Risikoforschung' in W. Krohn and G. Krücken (eds) *Riskante Technologien: Reflexion und Regulation. Einführung in die sozialwissenschaftliche Risikoforschung.* Frankfurt: Main, Suhrkamp, pp. 9-44.

Renn, O. and Zwick, M. (1997) *Risiko-und Technikakzeptanz.* Berlin: Heidelberg, Springer.

Sloterdijk, P. (1999) 'Regeln für den Menschenpark' *Ein Antwortschreiben zu Heideggers Brief über den Humanismus* (Special Print edition suhrkamp).

Thielemann, H. (1998) 'Kommunikation im Konflikt um die gentechnische Insulinherstellung bei de Hoechst AG' in O. Renn and J. Hampel (eds) *Kommunikation und Konflikt. Fallbeispiele aus der Chemie.* Würzburg, Koenigshausen and Neumann.

Torgersen, H. and Hampel, J. et al. (2002) 'Promise, problems and proxies: 25 years of European debate and regulation' in M. Bauer and G. Gaskell (eds) *Biotechnology – the making of a global controversy.* Cambridge: Cambridge University Press.

4 Predictive Medicine, Genetics and Schizophrenia

John Turney and Jill Turner

Introduction

It has often been suggested that more complete knowledge of human genes and gene sequences will make possible a new era of 'predictive medicine'. This is not a new idea (Harsanyi and Hutton, 1983). However, the emphasis on risk prevention has increased with the advent of the human genome project, and the future now envisaged depends on more detailed knowledge of a great many more genes at the molecular level (Hood, 1993).

Critiques of the bolder predictions associated with the HGP abound. Scientifically, it has been argued that such predictions rest on too narrow a view of aetiology, which in turn may derive from an outdated and excessively reductive notion of genetics (Strohman, 1993). Practically, some of the obstacles on the path to implementation of widespread testing have been thoughtfully discussed by Holtzman (Holtzman, 1989; Holtzman and Shapiro, 1998). And there has been much comment on the social influences on a wider culture of testing, and on its possible undesirable effects (Nelkin and Tancredi, 1989).

However, the principle of predictive medicine is still often invoked in discussions of future health policy and management (Bell, 1998; Genetics Research Advisory Group, 1995; Murray, 1996; Foresight Health and Life Sciences Panel, 1997). There is still a tendency to treat increased knowledge as an unproblematic good.

In this chapter, we explore how this assumption may prove problematic in particular contexts. We discuss the impact of possible future genetic findings in relation to a condition, schizophrenia, which has a contentious history, a complex aetiology and a high incidence. Schizophrenia has been an increasingly common subject of 'gene talk' in the last decade (Kitcher, 1996), but there are, as yet, no robust findings about particular gene loci. However, in the mid-1990s, workers in the field remained optimistic that such loci would be identified. A leading British researcher claimed in 1995 that 'we can be confident that, if genes of major effect are involved reasonably commonly in the aetiology of schizrophrenia, they will be detected and localized during the next few years' (McGuffin *et al.* 1995, p.681). In 1998, it was suggested by a US team reporting new findings concerning chromosome 13 that they had found 'the town, but not the street address' of a gene important in the development of schizophrenia (Connor, 1998).

Schizophrenia has also been frequently cited in more general discussions as a condition likely to be affected by new genetic knowledge. Such predictions have come from leading figures in the genome programme (Gilbert, 1993) and, more recently, from policy advisers. For example, a report to the British Department of Health in 1995 suggested that: 'It is estimated that 80-85% of the aetiology of schizophrenia and asthma is accounted for by genetic factors ... These discoveries offer the prospect of identifying at risk individuals' (Genetics Research Advisory Group, 1995).

What will we be able to do for these individuals who might be deemed 'at risk'? Some foresee major changes in management of a whole set of neurological and psychiatric disorders. Kaufmann, for example suggests that:

> The identification of major susceptibility genes for Alzheimers' disease, alcoholism, bipolar disorder, schizophrenia, and other psychiatric disorders can be expected to result in significant changes in our concepts of psychiatric nosology, etiology, and therapeutics.
>
> (Kaufmann *et al.* 1996, p.20)

There follows a strikingly well-articulated view of preventive possibilities:

> Regarding primary prevention, genetic discoveries may permit more informed genetic counselling, which historically has had to rely on unsatisfactory empirical risk data. This may allow realistic reproductive options to be exercised, especially in families heavily burdened by illness. Primary prevention may also involve gene therapy ...

> Secondary prevention may be achieved *prenatally*, in disorders like schizophrenia thought to develop in response to adverse *in utero* experiences. Pregnancies of foetuses at special genetic risk might be monitored more closely, and shielded from identified epigenetic risk factors like prenatal micronutrient deficiencies or viral exposures. Tertiary prevention might be achieved *postnatally* in schizophrenia and other disorders to the extent that high-risk presymptomatic individuals could be reliably identified and acute illness forestalled. Such interventions might be especially important if, as has been proposed, florid symptoms of both bipolar disorder and schizophrenia are pathogenic in their own right.
>
> (Kaufmann *et al.* 1996, pp.21-2)

The prospect outlined here must then include testing adults for the presence of certain alleles and informing them of the implications of the results. They will then, it is assumed, be advised how to reduce their risk of psychiatric illness. How exactly this might be done, and how such susceptible individuals might respond, is not discussed.

This is just one view of the prospects for uses of genetics in treatment of these disorders. Others are more sceptical, both of the prospects for clear-cut prediction, and about their possible consequences. The Nuffield Council for Bioethics, for example, suggests that genetic testing is unlikely to provide useful risk estimates for mental disorders. If it did, they suggest it would be necessary to consider questions of personal identity, anxiety, stigma and blame (Nuffield Council on Bioethics, 1998).

Together, these varied predictions suggest that it is important to ask how the world would look if there were known genes affecting schizophrenia. In particular, we report how the view varies between different professional and lay groups concerned with the condition in the UK: geneticists, psychiatrists, community psychiatric nurses, general practitioners, carers and persons diagnosed as having schizophrenia. Here, we report on the views of small numbers of individuals in each of these groups interviewed during 1996 and 1997. The results, we argue, show that the concerns expressed by the Nuffield Working Group are justified.

Background and methods

This chapter draws upon part of a 3-year research project funded by the Wellcome Trust. The project focussed on two common multifactorial diseases thought to have a genetic component: schizophrenia and heart disease. These were chosen to complement work already carried out or in progress on conditions like cystic fibrosis, breast cancer, muscular dystrophy and Huntington's disease (Marteau and Richards, 1996). Here, we discuss some of our findings in relation to schizophrenia.

This part of the study involved 39 semi-structured interviews with members of six separate 'interested' groups. These comprised six clinicians involved in genetic research, six psychiatrists, four community psychiatric nurses, five GPs, seven people diagnosed as having schizophrenia and 10 carers. Among the interviewees, one person understood to be diagnosed as schizophrenic turned out to be a manic depressive, and one terminated the interview shortly after it began. Both of these transcripts were discarded, to leave seven for analysis. The geneticists were experts in schizophrenia at various research centres in Britain and were identified through reviewing the scientific literature. The distinction between these two groups is not entirely clear-cut as two of the genetics researchers were also in clinical psychiatric practice, although none of those interviewed as psychiatrists was involved in research. The community psychiatric nurses were from a centre for promoting mental health in London. The remaining individuals were equally split between London and another university city and its outlying rural areas. The GPs and psychiatrists were recruited individually, while the carers and patients being treated for schizophrenia were recruited with the assistance of local officers of the National Schizophrenia Fellowship. None of the carers or schizophrenics were related to one another, and their experiences of the course of the condition were accordingly distinct.

The diagnosed schizophrenic patients and carers were the most varied groups. The people with schizophrenia were younger on average, between their early 20s and early 40s, and comprised three male and four female subjects. Three of the carers were young males, two with brothers and one with a father who had been diagnosed schizophrenic. The remaining seven were five women and two men who were middle-aged or elderly and had cared for a son or daughter diagnosed schizophrenic.

The interviews with professionals were typically shorter than those with other groups because of difficulties scheduling appointments, and those with GPs were all appreciably shorter – half an hour or less. However, it was still possible to elicit answers to the key questions posed to all the interviewees. Much of the interviewer's time with the carers and patients was taken up with extended rehearsal of their personal experiences, and these interviews frequently extended over 1.5-2 hours. As well as questions related to their particular perspective on schizophrenia, all respondents were asked for their views on the likelihood of genes of major effect being identified, the possible uses of such information – including the prospects for testing to identify 'at risk' individuals, and the kind of information people might need to help interpret any such test.

All the interviews, were audio-taped, coded for confidentiality and fully transcribed. The analysis of the data involved repeated individual scrutiny of the hard texts by the two researchers followed by a series of meetings to discuss the main themes arising. The themes and interview responses highlighted by each researcher were compared regularly, and relevant passages from each interview recompiled in separate text files for further consideration. The analysis was cumulative starting with initial findings of separate groups and progressing to consideration of the possible relationship of the different groups to each other. Fieldwork data for the second part of the study on heart disease fed into and informed our analysis.

The areas explored with each of the six groups of interviewees were similar, and were informed by some of the existing predictions in the literature, as detailed in the next section. Predictions about the genetics of schizophrenia are familiar to many people through the mass media (Connor, 1994). However, contemporary media reports, while they emphasize that research is under way, also generally recall that earlier announcements of genes implicated in schizophrenia and other behavioural disorders have typically been retracted. So our interviewees most often knew that genes and testing had been discussed in schizophrenia, but also knew that no such test was yet on offer. In each case, it was made clear that we were interested in being prospective–that is, that we recognized that no robust findings identifying any genes of major effect in the aetiology of schizophrenia have been reported, but wished to explore the consequences if this should happen. With the geneticists, the questions were more geared to eliciting their opinions about the likelihood of such findings emerging, as well as their views about the uses that might be made of them. With the other groups, we invited them to respond to the possibility of such findings, without necessarily commenting on whether they were to be expected.

Results

Our presentation of results begins with those closest to the state of the art in molecular biology, those researching the genetics of schizophrenia, and then moves progressively 'downstream', in terms of the production of certified scientific knowledge, discussing in turn the views of psychiatrists, general practitioners, community psychiatric nurses, close relatives as carers, and people with schizophrenia themselves. As we do so, the narratives offered to depict how new genetic knowledge might influence the experience of living with schizophrenia become increasingly complex.

When we invited interviewees to speculate on the prospects for a clearer understanding of the genetics of schizophrenia, responses varied widely.

The geneticists were generally clear that any gene located (and hence any test for such a gene) would be only part of the story. This is consistent with current published views of genetic investigators (Moldin and Gottesman, 1997). At its simplest, the gene might be a necessary but not sufficient condition for the onset of disease. As one geneticist put it, implicitly employing the metaphor of a precipitate threshold:

> It's always going to be a two factor model. You're going to have a gene, in these families, but you're going to have something else, to push it over into clinical illness.
>
> (Geneticist A)

This notion of two factors, translated into a simple two-step aetiology, was the most common way of summarizing the likely contribution of genetics. But it was typically accompanied by the suggestion that reality was likely to be more complicated.

The same interviewee, for example, stressed that the genetic side of the equation would itself turn out to be complex:

> with any gene that you find, you find most genes run in cascade. In other words they're only part of a sequence of events: one gene improves another gene, improves another gene, one protein another protein, another protein. And through the structure of that particular product that gene makes, we should identify a cascade of reactions, of which we can look at other proteins in that chain, which could be disrupted and causing susceptibility to schizophrenia in all the families.
>
> (Geneticist A)

Even this recognition of complexity, though, is expressed in a way that anticipates that there will be a single point of intervention. On the one hand, one gene or protein leads to many genes or proteins. On the other, when these in turn are examined, perhaps just one turns out to be increasing the schizophrenic risk 'in all the familes'.

Once such a gene or gene was identified, it ought to offer ways to help individuals at risk, possibly through 'more rational' drug therapy, or through prevention:

> Even with the illnesses which are supposed to be incurable, one thing we can do is we can identify the set at the beginning, the environmental factors because schizophrenia is not a disorder which will always present if you have a gene which is susceptible to schizophrenia. Identify the environmental factors, then you can modulate them, hopefully to prevent the illness ever expressing themselves. Okay?
>
> (Geneticist A)

Here the language suggests that the environmental factors being considered are of a relatively straightforward kind–exposure to drugs, let us say, rather than something more complex such as family or social circumstances. At least, it is not immediately clear how the latter might be 'modulated'.

Geneticists who foresaw the possibility of testing were conscious of some of the potential problems. They were aware of the pitfalls of a diagnostic-therapeutic gap, for example, and were hoping they would not arise in this case. As one said:

> Huntington's ... I suppose is the benchmark for genetic counselling, where we are counselling people, before they have the test, about dealing with this information, you know if they have the risk ... We're not in that situation with schizophrenia. I hope that when we are, we'll have something to offer those that are at risk.
>
> (Geneticist B)

They were convinced, though, that stressing hereditary influences on schizophrenia was helpful to patients and families. Perhaps counter-intuitively, a hereditary explanation was seen as de-stigmatizing, compared with notions of family pathology, which they perceived as clearly damaging, and were thankful were now outmoded. As one interviewee saw it:

> what you don't get so much any more is people thinking that they're to blame for their kids getting the disorder. So I think at least in the general public's mind the old concept of a schizophrenogic mother and all that is fortunately one.
>
> (Geneticist D)

Others, though, testified that these ideas were still encountered:

> I still have parents come up to me, um and wondering about their child, the upbringing of the person, if that had something to do with the fact that they had schizophrenia now.
>
> (Geneticist A)

When this occurred, in another researcher-clinician's view, it was largely due to the misguided ideas of other, less scientifically well-informed, professionals, and it was the geneticist's duty to put family members' minds at rest by emphasizing the more reassuring explanation:

> from a practical point of view ... you do meet people involved in social work or psychotherapy or family therapy who still think that schizophrenia isn't biological or they think no schizophrenia is genetic or family induced and that can cause a lot of misery to the relatives because they get blamed ... mothers or fathers get blamed for illness of their sons or daughters ... . Quite often when I see that happening the relatives actually believe it is biological and they are

> not to blame ... it does not take too much convincing ... they are relieved when I tell them they are not responsible and I have long given up any pretence of saying well they might be you know ... might be right to some extent.
>
> (Geneticist F)

The genetics researchers, then, were on the whole confident that there would be stronger genetic explanations of schizophrenia, and that these explanations would lead to beneficial applications. But even without the latter, they saw benefits in the genetic explanation in and of itself. The responses from other groups concerned with schizophrenia give less reason for confidence about these benefits. It was rarely, if ever, a case of denying them outright, more of raising unacknowledged complexities and ambiguities, both in weighing the implications of genetic information and in its application.

Other professionals, for example, saw the clear preference for a genetic explanation as too simple a view, their own professional practice producing a different perception about what kinds of discussion to have with whom. As one psychiatrist suggested, the level and nature of interest in possible causes of the condition was

> very much dependent on what kind of cultural group you're talking about, and sort of socio-economic group. The less educated people tend to be less inquisitive. Whereas working in H. people are much, much more interested in the background of the illness they have, because they tend to be a more educated, inquisitive population. And certain cultural groups have a different slant on mental illness as well and, some groups are very much keen that it is very biological and all this, and that it comes from this and there's nothing they can do about it and they very much see that as removing themselves from any kind of blame. Because unfortunately with a lot of mental illness there is a lot of stigma and blame attached to it. Um, whereas other groups they would look at more sort of socio-environmental factors to be causative. Again to try and attribute a cause to it.
>
> (Psychiatrist A)

As well as this recognition that both the degree of interest in an explanation and the kind of explanation that made most sense varied between groups, there was also emphasis that the consideration of causes was strongly influenced by personal histories. As another psychiatrist related:

> Particularly for mental health problems people are keen to try and work out what sort of, what went wrong really. Parents are often concerned more from a genetic point of view, "is it something we pass on with genes?" I've had people ask me "is it genetic?", but people also ask, and they have their own reasons, "could it be due to him being born prematurely" or "him taking drugs at school?" or whatever, so they come with a question "could this be at least partly due to this?" and then often that leads on to a discussion of other causes or other factors involved.
>
> (Psychiatrist D)

The psychiatrists were also notably wary about the possibility of any kind of predictive test for people who might be at risk for schizophrenia. They anticipated demand for such a test as soon as it was publicized:

> I would need to be reassured about the sensitivity and specificity of the test to be involved I guess in using it. Though I can see inevitably the minute such things are postulated that people are keen to get involved. There are always people, you know, who say they want to know.
>
> (Psychiatrist D)

Another respondent envisaged a struggle to avoid misinterpretation of any such test:

> I find them very frightening ... they need to be used in conjunction with, you know, a great deal of counselling and you need to have a great deal of knowledge as well. I mean, okay, if they did come out and they can formally predict, you know, who is and isn't at risk of schizophrenia then so be it. It may be that then we can develop counselling and explain exactly what those results mean to people. I suspect it will be extremely complicated and in great danger of um, being 'bucked' up by the media and er, and misunderstood so that people don't really know what's being tested and what the results of the rest mean.
>
> (Psychiatrist B)

There is concern here about the difficulties of working with individuals if testing becomes possible. Even if geneticists appreciate the complex, multifactorial nature of the condition, this will not be reflected in media reports, it is suggested, and will have to be painstakingly reintroduced in counselling sessions.

The 'frightening' aspect of testing referred to here, though, is to do with wider social effects of testing, and is linked to possible eugenic consequences of testing. Concerns here focused on possible ante-natal tests:

> the idea that people think that they can go and have a test, either it's an ante-natal test or whatever for schizophrenia or whatever, and start saying "oh well it looks like this baby might have schizophrenia", you know, "get rid of it", what does that actually say about mental illness and attitudes towards people with schizophrenia?
>
> (Psychiatrist B)

This rhetorical question shows a readiness in this group to link genetics to eugenics, which is largely absent from the responses of the geneticists, who as others report generally draw discursive boundaries in ways that keep the two separate (Kerr, Cunningham-Burley and Amos, 1997).

Finally, in line with a conviction that different explanations mean different things to different people, the psychiatrists were more inclined than the geneticists to have mixed feelings about the consequences of testing in adults. They saw the sense in early warning and prevention, if possible, but were wary of the adverse effects of bestowing information like this on families. As one explained:

> from a treatment point of view, if we knew which people were vulnerable we could keep a closer eye on them and if there are any signs we could treat them earlier, maybe even prevent developing the full blown illness. So that would be a positive side. On the negative side, I mean it would be disastrous for some families, who already probably have increased expressed emotion. There'd be enormous blame attributed to people. I mean it's horrendous in Huntington's families that I've been involved with: the family dynamics there, and the recriminations that a person's brought their genes into that family.
>
> (Psychiatrist A)

While this comment appears to stem from experience with another disease, it at least suggests that genetic explanations are not automatically de-stigmatizing or solvents for guilt or shame.

The psychiatrists, then, conventionally the highest status 'experts' about schizophrenia, already had reservations about the potential effects of using new genetic information which were not raised by their colleagues researching in genetics. These reservations were shared, to a greater or lesser extent, by members of all the other groups investigated, though inflected in different ways by their varied professional training and experience and their personal histories.

The four community psychiatric nurses interviewed raised a similar set of concerns to the psychiatrists, but often expressed in stronger terms. In particular, they were opposed to any tendency to elevate genetics to what one might call the dominant aetiological narrative. As one explained it:

> I'm quite open actually, I hate labelling people. It's easy to say is it because it's in your family, it's quite easy to say that. But sometimes the evidence is staring you in the face really, and you sometimes have doubts about it so my mind's not completely closed. I think there are other factors as well.
>
> (CPN A)

For this interviewee, keeping open the question of genetics versus environment was a virtue, a matter of avoiding 'labelling' – a negative term for some in this group – which was seen as more strongly implied by a genetic than an environmental cause, presumably because the gene, once identified, is inescapable.

She made clear that this stemmed in part from her own experience of a genetic condition. "It's another way of labelling, selecting this particular group, this particular gene. It's the same thing, every time I go to hospital for example, I get tested for sickle-cell, and the way that feels for me" (CPN A).

Their experience as the primary official carers for many long-term schizophrenic clients also gives the CPNs a well-elaborated idea of what 'environment' means, in social terms rather than the biological terms (intra-uterine environment or viral insults), typically mentioned by geneticists.

> I mean when I've actually worked in places like B. where the clientele is quite different, they've got more privileges and things like that, and the standard of living is much better. I didn't think it was so marked, like here there's a housing problem and social problems, unemployment all sorts of other problems and I think those play a part as well. And if all the members of the family are subjected to that ...
>
> (CPN A)

As with the psychiatrists, the CPNs did not necessarily accept the geneticists' assumption that a hereditary explanation would be a relief from blame. However, like most groups aside from the geneticists, there was no unanimity about this or other aspects of the genetic prospects for schizophrenia. One CPN, for instance, endorsed the argument that a genetic story was preferable to earlier, more distressing explanations, and saw other possible benefits in terms of attitudes to the condition:

> In a sense it can cut both ways but as far as having, I mean the blame aspect, I think a lot of lay people would say: "well you need to pull yourself together or distract yourself". You know you can't see schizophrenia in as much as you can see a broken leg. So if you seem distressed you can't see it. It's something you'll be able to shake off or get on with. Like if they did something different that would make a difference like if they had a better hobby or job then their problems would go away. People take that on board, so having a label as such can help people to understand. They can say: "well, thank God it's not my fault".
>
> (CPN B)

The argument is analogous to the position of those who believe that belief in a 'gay gene' will enhance tolerance for male homosexuals because no blame can attach to the individual for something that is inherited (Turney, 1998). But the same respondent also emphasized the possible drawbacks:

> The down side is I suppose it can make people feel quite hopeless and out of control. You know: that's it, I've no chance, I can't get away from it, it's there for ever. Or deny it and say: "well I haven't got something wrong with me. This diagnosis, that's just a label you've given me, I'm not going to accept it therefore I'm not going to accept your help or your medicine".
>
> (CPN B)

Here the interviewee's speculation includes the alternative of fatalism or activism discussed by others in relation to predictive testing for, say, heart disease (Davison, 1996). But there is an additional possibility canvassed, active opposition to the genetically grounded diagnosis.

The same interviewee also expressed concern about future relations between this group of professionals and their principal clients:

> I think if somebody said "it's a genetic disorder" ... it would make it harder for, it would create a more "us and them" situation. And I think the repercussions for us, I mean I feel it now as a nurse, we've become more and more agents of social control, people's lives aren't their own any more to a certain extent ... I think if that happened tomorrow ... that would reinforce that. I think that our client group would start to feel that we were more like policemen than nurses.
>
> (CPN B)

These CPN's views were typically shaped by experience of working with large caseloads of clients with schizophrenia. GPs handle a much more diverse patient group, and have proportionately much less experience of schizophrenia. Nevertheless, in our group of GPs, their more diverse experience did not make them any more optimistic about the benefits of a more refined genetics of schizophrenia. Some had deep reservations about the general suggestion of predictive medicine. They questioned whether all knowledge was worth having, as in the following extract:

> I know this business of refining the genes, we can screen the patients regularly and make sure they are not carrying the disease and so on, and of course all that sounds very good and worthwhile but, what must it be like to be born and know that you have a gene that might well influence your development sufficiently so that you develop schizophrenia or breast cancer or whatever else there is? I mean do you really want to live under that shadow? My instinct is to say, no we don't, as human beings it will do us nothing but harm.
>
> (GP C)

The carers group, who had seen close relatives trying to cope with schizophrenia, were also conscious of this problem. Carers were unanimously in favour of *pre-natal* testing if it permitted choice whether to bear a child with a susceptibility to schizophrenia. They wanted other family members to be spared their own extremely harrowing experiences. They had reservations, though, about testing with a view to prevention during child-rearing.

On the one hand, there would be sense in acting on genetic risk information:

> I think if I were to have children knowing that there was a chance that they would go on to develop this disorder but then also being aware that there may be things in the environment that could contribute to them developing that

> disorder, then one can perhaps take positive steps to altering your behaviour, so that you reduce the chances of them developing the disorder.
>
> (Carer D)

On the other hand:

> That having been said, you'd perhaps then go through a period of child rearing almost looking for something wrong and always wondering, are they going to get it, are they going to get it, and if they do then having that, the additional worry it's because I didn't do this (INDISTINCT).
>
> (Carer D)

As far as adults already diagnosed with schizophrenia were concerned, carers saw little point in discussing genetic explanations. It was, perhaps, too late to be bothered either way:

> I don't know how it ... I mean if somebody said to S., the reason why you've got this trouble, so to speak, is because you've got something wrong with your genes or so, so you might say "that's very interesting thank you very much". But she'd just carry on living her life wouldn't she? It wouldn't make any difference?
>
> (Carer G)

Most of the group diagnosed with schizophrenia agreed. However, they, too, could see some merit in predictive testing. A number had ruled out having children, but those who had not tended to be fearful:

> I'm wondering if it's going to happen to my kids, you know?
>
> (Schizophrenic E)

One interviewee had excluded having children on genetic grounds because of the apparent pattern of illness in her family:

> I've just known that I wouldn't have children. Because I mean in our family it's just so obvious that it is a genetic thing you know. It may not be in everybody's case but, it, it's just absolutely obvious that it is here.

She foresaw strong family interest in any future test:

> if the test was fairly safe I would probably have a test! I'm sure everybody in our family would have a test you know ... I think they'd be straight down the hospital with their kids.
>
> (Schizophrenic A)

This interviewee was one of several who were deeply interested in patterns of inheritance because of their family history. Even so, this did not necessarily engender confidence in any ultimate explanation. One such subject, who also reported that his father was an alcoholic wife-batterer and that he himself had been a multiple drug user, eventually concluded that:

> I'ts too complicated for someone like me to understand you know, all I know is I hear voices, I get delusions, I'm ill, and I've got to make the most of what I know about.
>
> (Schizophrenic B)

To complicate matters further, this fatalistic stance was combined with an apparent faith in the eventual success of science, a success in which the search for genes might well play a part. While the immediate personal relevance of any new genetic information would be minimal:

> I think they will manage to find a cure in the end, cos science just carries on 24 hours a day, 365 days a year, so I mean, as long as they stick at it and they get funding, I don't see why they shouldn't be able to cure at the end of the day, but I don't know when. And if they find the sets of genes, yeh that, that might help the process, but to the everyday lay schizophrenic who's just on a ward, it won't make much difference to them. Obviously I understand my dad was a schizophrenic, fair enough, what's it got to do with me you know?
>
> (Schizophrenic B)

Here we see a close similarity between the simplest explanation of the geneticists, that finding genes ought to be helpful in improving treatment, and the cautiously hopeful view of a young man actually experiencing the disease. But the passage also exemplifies the need of people with schizophrenia to construct a narrative that makes sense of their lives (Barham, 1993). For this interviewee, the narrative of the onward movement of genetic science had no immediate connection with his personal narrative.

One aspect of the context that emerged from the discussion of prospects for raising children was the extent to which those deemed mentally ill mix with others similarly diagnosed. This raised quasi-eugenic considerations for some of this group, one of whom offered an elaborate testing scenario in which there would be advantage if two schizophrenics wanted to have children, because they could take eggs and sperm and run DNA tests on them to select gametes that would produce a child without schizophrenia:

> Because I think that's quite a major worry. For people who've been mentally ill, whether it be one parent or both, you know if they have a child through the natural process ... Would it have mental health problems when it's very young, which would be tragic I think.
>
> (Schizophrenic C)

Finally, one of this group endorsed the idea that predictive testing might be of use to parents who wanted to try and protect their offspring from pressures that someone at risk for schizophrenia should be spared. Again, this conclusion drew on a personal narrative:

> well it [i.e. testing] would be useful because ... I think you could try and lessen the chances of someone having full-blown schizophrenia. You could avoid, I mean for example when you choose what school they go to. I mean whether they go to a great high powered school or something more middle of the road, or something low key. I mean that can make all the difference between having a breakdown or not, if someone's got schizophrenic genes. I mean it could have a vast impact on someone's life, I think if you could do a test, do a genetic test on people who may be carrying the gene.
>
> (Schizophrenic C)

This interviewee acknowledged that there might be some problems if testing were used in this way, notably a risk of stigmatizing the child, but was still convinced about the balance of risk and benefit:

> I think the argument in favour of testing far outweighs the arguments against.
>
> (Schizophrenic C)

The group diagnosed with schizophrenia, then, seem more inclined to see advantage in a genetically based diagnosis or prediction than the non-geneticist professionals. But they by no means see the implications of improved genetic knowledge as straightforward. They convey a strong sense of struggling to make sense of lives that incorporate experiences far out of the ordinary. A number were clearly fascinated with questions of genetics and heredity, both in terms of accounting for their own experience and helping them resolve questions about possible future child-rearing. But others were equally strongly of the view that, even if improvements in genetic understanding might be desirable, they would have no bearing on their personal situation. More than any of the other groups, they seem simultaneously to share some features of the geneticists' outlook, but also to be worlds away from them. This is what should affect our thinking about the consequences of advances in molecular genetics yielding results that would bind the lives and practices of all these groups more closely together.

Discussion

Our small-scale study suggests that in-depth interviews reveal a diversity of views about the likely implications of improved genetic knowledge about schizophrenia across the six groups we investigated. Our results, though we cannot claim with any certainty that they are representative, do cover a wider range of groups than earlier studies documenting lay and professional ideas about the origins of schizophrenia (Wahl, 1987; Furnham and Rees, 1988) and lay ideas about genetic risk (Parsons and Atkinson, 1992). They also extend our knowledge of possible psychological and social consequences of genetic testing to a condition not previously considered in great depth (Shiloh, 1996; Davison, Macintyre and Davey Smith, 1994; Marteau and Croyle, 1998). Dealing with schizophrenia meant that we were deliberately inviting our respondents to speculate, so there is no assurance that their actual behaviour would match their suggestions here if new genetic information became available. Experience with Huntington's disease, for example, suggests strongly that declared intentions about testing are not borne out in practice (Spijker and Kroode, 1997).

However, differences between the groups do suggest that significant difficulties and ambiguities will complicate the prospects foreseen by the Genetics Research Advisory Group or by writers such as Kaufmann, as quoted in the introduction. Our results indicate that views about risk reduction on the basis of genetic information must be considered in relation to complex histories of family struggle to cope with the disease, and the way these histories are perceived by a range of professional and lay groups. This will influence the response to new genetic information in contradictory ways.

It does appear, for example, that there may be a strong demand for genetic counselling if genes of major effect are identified in schizophrenia. Those with schizophrenia, carers and other family members indicated that they will take note of such claims, even if the reported associations are relatively weak. While expressed intention in response to a hypothetical question may mislead, we note that there is already a relatively strong awareness that the condition may run in families, even if this is not expressed in genetic terms alone (Jorm *et al.* 1997). The wide publicity which genetic findings are sure to attract may well crystallize a concern about future prospects.

The demand is likely to be of two kinds, in the main – from first degree relatives (siblings and children) of people with schozophrenia, who may seek testing to clarify their own susceptibilities, and from people with schizophrenia or relatives who want advice on reproductive plans. There will also be queries from prospective adoptive parents of children with a family history of schizophrenia (Office of Technology Assessment, 1994). It is not at all clear, incidentally, who might be willing or able to offer any necessary counselling – which is likely to demand an unusual combination of genetic and psychiatric expertise (Schulz, 1982). This point is also made by some who gave evidence to the Nuffield inquiry already cited (Nuffield Council on Bioethics, 1998; see also Kinmouth, 1998).

But in spite of this apparent willingness to make use of testing, there is considerable scepticism about its use. Some professionals see pre-natal testing for susceptibility to mental disease, in combination with hostile or fearful public attitudes (Rabkin, 1974) as a step toward negative eugenics. As Kerr *et al.* have shown, though, geneticists usually draw a clear line between a eugenic past and a future of individuals benefiting from new genetic information (Kerr *et al.* 1997).

For post-natal testing and prevention, it is currently hard to see how an individual or a family might use information suggesting a greater-than-average risk of schizophrenia. Unless and until there are more effective therapies, the best advice may remain general avoidance of stress or emotional strain. How this is to be translated into day to day family management seems deeply problematic. There is certainly a chance that such advice would increase stress rather than reduce it. As families caring for a schizophrenic individual are already under enormous stress, this might be the worst of all possible worlds. And although genetic explanations may reduce feelings of guilt, they may also increase it, if prevention 'fails'.

As with guilt, so with stigma. In general, the literature suggests that questions of stigma, responsibility and identity will be played out differently for different conditions, depending on the social and cultural history of each one. In addition, they may be played out differently in relation to a particular condition, depending on the social location and experience of the individuals involved (Davison *et al.* 1994; Davison, 1996).

For example, some of our respondents did endorse the belief that there will be a de-stigmatizing effect if a genetic explanation of schizophrenia gains ascendancy. Some people will see the passing on of a gene or genes as less 'their fault' than implied by earlier aetiologic ideas like the pathological family. But this will not be universal, and other families will undoubtedly see the suggestion of 'bad blood' associated with a heritable component of schizophrenia as highly stigmatizing. Hard and fast predictions here are and will remain extremely difficult, but the evidence of wide variation even within our small sample suggests they will depend on knowing the individuals concerned.

Related to stigma, and to questions of fatalism and personal responsibility, there are troubling questions of labelling that are likely to emerge as the genetics of schizophrenia develop. At present, schizophrenic individuals have a wide range of attitudes to the diagnostic label for their condition (Link, 1987). Among our respondents, some deny its validity, seeing it as a psychiatric conspiracy, some see it as racist, some accept it with reservations, and some see it as a useful device for securing support from various sources. Many share the attitude that the label is, in some degree, negotiable. They may well feel that this negotiability is lost if the diagnosis is backed with a genetic warrant. Some will then conclude that their fate is inescapable, and that science has painted them into a diagnostic corner. Given the wide range of schizophrenic histories, from a single episode followed by complete recovery to chronic recurrence, this would be unfortunate. This alone is a strong argument for great care in the presentation of any new findings on the genetics of schizophrenia.

That presentation, whether by molecular geneticists to clinicians, newspapers to the public, or health professionals to their patients or clients, is marked by the tension between a desire to acknowledge the multifactorial aetiology of the condition and a tendency to revert to an implied two factor model – 'genes plus environment' – for ease of explanation. The complexity of the contexts in which symptoms classified as schizophrenic arise speaks strongly against simple metaphors of causation as a guide to action. A genetic influence on susceptibility to schizophrenia, if there is one, is not best understood as a 'trigger', or any other simple mechanical or chemical metaphor. It is one consequence of a whole complex of life events, which in turn leads to others, unfolding over time (Bateson and Martin, 1999). It would be encouraging to believe that if we can test for the presence of genetic factors before a disease strikes, we can warn 'those who are susceptible to stay away from specific environmental triggers', but there is little reason to suppose that the lives of those who develop schizophrenia will lend themselves to such prescription.

Acknowledgements

Supported by Wellcome Trust grant number 042876. We thank Adam Hedgecoe, Brian Balmer, Helen Lambert and members of the London Public Understanding of Science seminar for discussion and advice on the manuscript.

References

Barham, P. (1993) *Schizophrenia and Human Value*, London: Free Association Books.

Bateson, P. and Martin, P. (1999) *Design for a Life: How Behaviour Develops*, London: Jonathan Cape.

Bell, J. (1998) The new genetics in clinical practice, *British Medical Journal*, 316, 21 February, pp. 618-20.

Connor, S. (1994) Deep waters in the gene pool, *The Independent on Sunday Magazine*, 13 November, p. 70.

Connor, S. (1998) Schizophrenia gene close to discovery, say scientists, *The Independent*, 1 September, p. 4.

Davison, C. (1996) Predictive genetics: the cultural implications of supplying probable futures in T. Marteau and M. Richards (eds) *The Troubled Helix: Social and Psychological Implications of the New Human Genetics*, Cambridge: Cambridge University Press, pp. 317-30.

Davidson, C. (1997) Everyday ideas of inheritance and health in Britain: implications for genetic testing in A. Clarke and E. Parsons (eds) *Culture, Kinship and Genes*, Chapter 12, Macmillan: Wellcome.

Davison, C., Macintyre, S. and Davey Smith, G. (1994), The potential social impact of predictive genetic testing for susceptibility to common chronic diseases – review and research agenda, *Sociology of Health and Illness*, 16, pp. 340-71.

Foresight Health and Life Sciences Panel (1997) *Progress through Partnership*, London: HMSO.

Furnham, A. and Rees, J. (1988) Lay theories of schizophrenia. *International Journal of Social Psychiatry*, 34, pp. 212-20.

Genetics Research Advisory Group (1995) *A First Report to the NHS Central Research and Development Committee on the New Genetics*, London: Department of Health.

Gilbert, W. (1993) 'A Vision of the Grail', in D. Kevles and L. Hood (eds) *The Code of Codes: Scientific and Social Issues in The Human Genome Project*, Cambridge, MA: Harvard University Press, pp. 83-97.

Gottesman, I. (1996) 'Blind Men and Elephants – Genetic and Other Perspectives on Schizophrenia' in L. Hall (ed.) *Genetics and Mental Illness – Evolving Issues for Research and Society*, New York: Plenum.

Harsanyi, Z. and Hutton, R. (1983) *Genetic Prophecy – Beyond the Double Helix*, London: Granada.

Holtzman, N. (1989) *Proceed With Caution: Predicting Genetic Risks in the Recombinant DNA Era*, Baltimore, MD: The Johns Hopkins University Press.

Holtzman, N. and Shapiro, D. (1998) The new genetics – genetic testing and public policy, *British Medical Journal*, 316, 14 March, pp. 852-6.

Hood, L. (1993) 'Biology and Medicine in the Twenty-First Century' in D. Kevles and L. Hood (eds) *The Code of Codes: Scientific and Social Issues in the*

Human Genome Project, Cambridge, MA: Harvard University Press, pp. 136-63.

Jorm, A., Korten, E., Christense, H., Rodgers, B. and Pollitt, P. (1997) Public beliefs about causes and risk factors for depression and schizophrenia, *Social Psychiatry and Psychiatric Epidemiology*, 32, pp. 143-8.

Kaufmann, C., Johnson, J. and Pardes, H. (1996) Evolution and revolution in psychiatric genetics in L. Hall (ed.) *Genetics and Mental Illness – Evolving Issues for Research and Society*, New York: Plenum.

Kerr, A., Cunningham-Burley, S. and Amos, A. (1997) The new genetics: professionals' discursive boundaries, *Sociological Review*, 45, pp. 279-303.

Kinmouth, A. (1998) Implications for clinical services in Britain and the United States, *British Medical Journal*, 316, 7 March, pp. 767-70.

Kitcher, P. (1996) *The Lives to Come: The Genetic Revolution and Human Possibilities*, London: Allen Lane.

Link, B. (1987) Understanding labelling effects in the area of mental disorders: an assessment of the effects of expectations of rejection, *American Sociological Reviews*, 52, pp. 96-112.

Lupton, D. (1995) Taming uncertainty: risk and diagnostic testing, *The Imperative of Health – Public Health and the Regulated Body*, London, Berkeley: Sage.

Marteau, T. and Croyle, R. (1998) Psychological responses to genetic testing, *British Medical Journal*, 316, pp. 693-6.

Marteau, T. and Richards, M. (eds) (1997) *The Troubled Helix: Social and Psychological Implications of the New Genetics*, Cambridge: Cambridge University Press.

McGuffin, P., Owen, M. and Farmer, A. (1995) Genetic basis of schizophrenia, *The Lancet*, 346, 9 September, pp. 678-82.

Moldin, S. and Gottesman, I. (1997) Genes, experience and chance in schizophrenia – positioning for the 21st century, *Schizophrenia Bulletin*, 23, pp. 547-61.

Murray, T. H. *et al.* (eds) (1996) *The Human Genome Project and the Future of Health Care*, Bloomington: Indiana University Press.

Nelkin, D. and Tancredi, L. (1989) *Dangerous Diagnostics: The Social Power of Biological Information*, New York: Basic Books.

Nuffield Council on Bioethics (1998) *Mental Disorders and Genetics: the Ethical Context*, London: Nuffield Council on Bioethics.

Office of Technology Assessment (1992) *The Biology of Mental Disorders*, Washington, DC: US Government Printing Office.

Office of Technology Assessment (1994) *Mental Disorders and Genetics: Bridging the Gap Between Research and Society*, Washington, DC: US Government Printing Office.

Parsons, E. and Atkinson, P. (1992) Lay constructions of genetic risk, *Sociology of Health and Illness*, 14, pp. 437-55.

Rabkin, J. (1974) Public attitudes toward mental illness: a review of the literature, *Schizophrenia Bulletin*, 10, pp. 9-33.

Schulz, P. (1982) Patient and family attitudes about schizophrenia: implications for genetic counselling, *Schizophrenia Bulletin*, 8, pp. 504-13.

Shiloh, S. (1996) Decision-making in the context of genetic risk in T. Marteau and M. Richards (eds) *The Troubled Helix: Social and Psychological Implications of the New Genetics*, Cambridge: Cambridge University Press, pp. 82-103.

Spijker, A. and Kroode, H. (1997) Psychological aspects of genetic counselling: a review of the experience of Huntington's disease, *Patient Education and Counselling*, 32, pp. 33-40.

Strohman, R. (1993) Ancient genomes, wise bodies, unhealthy people: limits of a genetic paradigm in biology and medicine, *Perspectives in Biology and Medicine*, 37 (1), pp. 112-45.

Turney, J. (1998) The 'gay gene' in science and the media. *Open University S802, Science and the Public. Part A Case study. SUP392178*, Milton Keynes: Open University.

Wahl, O. (1987) Public versus professional conceptions of schizophrenia, *Journal of Community Psychology*, 15, pp. 285-91.

PART II
COMMERCIALISATION AND HEALTH

5 Pharmacogenetics: Implications for Drug Development, Patients and Society

Alun McCarthy

Introduction

Until recently, clinical genetics was confined to the study of rare and severe diseases such as Huntington's disease and cystic fibrosis, which are caused by a single gene mutation and inherited in a simple Mendelian fashion. Advances in genetic science and technology have now made it possible to begin to understand the genetic basis of common diseases which are more complicated to study owing to the number of genes involved, the complex inheritance patterns and the environmental factors also involved.

Today, human genetics has a high media profile, and research findings are increasingly being applied to clinical practice. There is a significant research commitment to the field, not only in academia but also in the pharmaceutical industry. Genetic research at Glaxo Wellcome has two principle goals in terms of healthcare provision: firstly, to enhance the understanding of common diseases with unmet medical needs, so that new targets can be discovered and new treatments made available and, secondly, to understand the genetic basis of patients' responses to our medicines, such that treatment can be targeted to those likely to gain most benefit (pharmacogenetics).

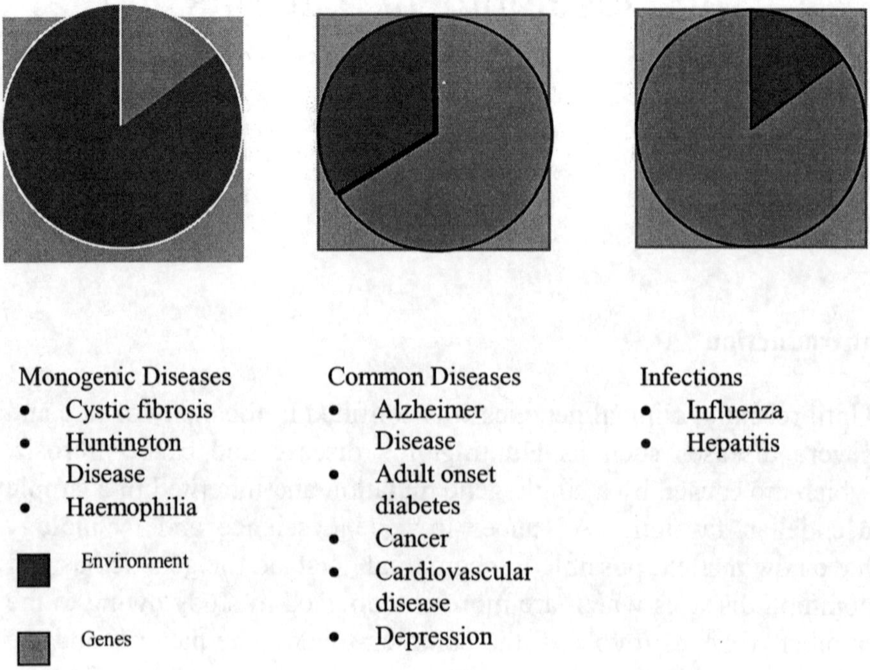

Figure 5.1 Genes and disease

Genes and disease

We can divide the associations of genes and disease into three main groups: monogenic diseases, common complex diseases and infectious diseases (Figure 5.1).

The term 'monogenic diseases' covers those severe but relatively rare diseases that are caused by the expression of a different form of a single gene. These diseases are sometimes referred to as Mendelian disorders, as they follow inheritance patterns described by Gregor Mendel over 100 years ago. More than 4000 diseases are thought to be single-gene disorders, and well-known examples are cystic fibrosis, Huntington's disease and haemophilia. The most important factor for the manifestation of the disease is presence of the mutated gene: other environmental factors play a much smaller role.

In the common complex diseases that affect many millions of people (such as Alzheimer's disease, adult onset diabetes, cancer, cardiovascular disease and depression), there is often an important genetic component, which is due to a number of genes that alter the risk of the disease occurring. These are called susceptibility genes. For these diseases, often referred to as polygenic diseases, the environment has a greater impact than in the case of the monogenic diseases. Research has indicated that the aetiology of most, if not all, of these common diseases is significantly impacted by genetic variation in the population.

In the third group of diseases, there may be a genetic component to an individual's susceptibility, but the environment has a much greater influence. The obvious examples of these illnesses are the infectious diseases, such as influenza and hepatitis, that arise following exposure to a specific pathogen in the environment.

Common complex diseases

One frequently hears someone say that a common complex disease, such as asthma, 'runs in our family'. Typically, there is a family history of affected individuals but no clear pattern of inheritance. An individual's risk of disease increases with a positive family history.

The common complex diseases affect many millions of people world-wide and result from the interaction of 'susceptibility genes' with environmental factors. In addition to the examples already cited, the group of diseases includes arthritis, asthma, obesity and uni/bipolar depressive disorders. The diseases have major medical and patient impact and significant unmet medical needs.

Susceptibility genes confer an altered risk for a disease but, alone, are not usually sufficient to cause that disease. Susceptibility genes may influence the age of onset, contribute to the rate of disease progression or help to protect against the disease.

An example is the APOE gene on chromosome 19, which codes for apolipo-protein E, and is a susceptibility gene for late-onset Alzheimer's disease. Three common forms of APOE-APOE2, APOE3 and APOE4-are distributed throughout the human population. APOE4 increases the risk of Alzheimer's disease and lowers the age of onset, whereas APOE2 seems to offer protection against the disease. The genetype at the APOE locus (e.g. E4/E4, E3/E4 etc.) can be used to increase the accuracy of diagnosing Alzheimer's disease in an individual with impaired cognitive function.

By identifying disease susceptibility genes and the role of the different forms of the genes in disease processes, new insights will be generated that greatly increase our understanding of the causes of these diseases. For example, APOE is well known as a plasma protein involved in circulating lipids in the bloodstream: its role as a susceptibility gene in Alzheimer's disease has opened up completely new ways of studying and understanding the pathophysiology of this disease. These new avenues can then lead to compounds which act at the root cause of these diseases. This will lead to the development of new medicines which can stop or prevent disease.

Pharmacogenetics

Currently, many medicines are prescribed through a trial and error process. The patient presents with an illness, a diagnosis is made, and a medicine and dose are selected based upon the mean response of patients in clinical trials and the physician's own experience. There is effectively a trial period and subsequent follow-up to determine whether an effective response with tolerable side effects is achieved. If an initial dose elicits no or insufficient response, progressively higher doses may be tried until an effective response is obtained, or another medicine may be prescribed instead of-or as well as-the first. If side effects are intolerable, a lower dose may be appropriate or the medicine may need to be changed. A delay in achieving a therapeutic response can be worrying or cause complications for the patient and can also result in poor patient compliance as patients gain no benefit from taking the initially prescribed medicine.

Several factors, which may vary between individuals, influence a patient's response to treatment including:

- amount and rate of medicine absorption;
- rate of drug metabolism and elimination;
- drug concentration at the drug target;
- variation at the drug target, e.g. differences in the number of receptors or receptor morphology;
- second messenger mechanisms.

Genes coding for proteins that modulate these processes are known to vary in the population, and we can expect that these genetic variations affect an individual's response to a medicine. Measurement of these genetic variants in patients will enable patients' responses to treatment to be predicted and, hence, appropriate therapy targeted to the individual (Anderson *et al.* 1999). The parameters listed above are most likely to impact the efficacy of a medicine in an individual: the same argument can be made about other genetic variants that may impact possible adverse reactions that may arise from use of medicines. We envisage that, for example, a patient will be diagnosed with an illness for which two different medications-medicine X and medicine Y-are available. Using the patient's genetic make-up, a simple medicine response profile might predict that medicine X has a 90% chance of being effective but there is a 2% risk of severe liver failure, while for medicine Y there is a 75% chance of effectiveness and virtually no risk of any serious side effects.

There are an increasing number of example in the literature where patients' genotype is correlated with their response to a medicine. Kuivenhoven *et al.* (1998) showed that variations in the gene that encodes the cholesteryl ester transfer protein (CETP), which is involved in the metabolism of high-density lipoprotein, influences the response to pravastatin. Their results showed that the change in mean luminal arterial diameter (a measure of atherosclerotic disease) in response to pravastatin/placebo treatment was related to the CETP TaqIB genoptype. With the B1B1 genotype, there was a clear response to prevastatin compared with placebo but B2B2 carriers did not appear to benefit from pravastatin treatment.

The work of Tan *et al.* (1997) shows how variations in the drug target or receptor can influence a patient's response to treatment. Polymorphisms at codon 15 and 27 of the B2-adrenoceptor gene are quite common in asthmatic and non-asthmatic people alike. Tan *et al.* studied 22 moderately severe stable asthmatics who were treated with inhaled placebo and inhaled formoterol for 4 weeks each in a crossover study. Although the study was small, the results indicated that there may be a subset of patients-those with a particular form of the B2-adrenoceptor-in whom bronchodilator desensitization occurs when formoterol is used on a regular basis. The maximum forced expiratory volume (FEV_1) showed a significantly greater degree of bronchodilator desensitization in subjects with the homozygous Gly-16 form of the receptor than in those with the heterozygous Arg-16/Gly-16 or the homozygous Arg-16/Arg-16 form.

The polymorphisms in cytochrome P450 isoenzymes that are involved in the oxidative metabolism of many medicines are relatively well known. For example, the cytochrome P450 isoenzyme, CYP2D6, is known to be involved in the metabolism of over 40 medicines. There are different variants of the isoenzyme, some of which have impaired function. Individuals who have a dysfunctional isoenzyme variant are called poor metabolisers and can respond differently to medicines that rely on this route of metabolism.

Pharmacogenetics and drug development

At Glaxo Wellcome, we are trying to identify the alleles/genetic markers involved in clinical outcome by correlating patients' genotype with their medicine-related phenotype. The goal is the delivery of the right medicine to the right patient, and the majority of clinical research programmes of investigational medicines include genetic research. Current technology enables variations in specific genes to be correlated with patient outcome. Therefore, hypotheses have to be generated relating to which genes may be involved in the drug's metabolism and mode of action. Where associations are found, further studies will need to be conducted to confirm the finding.

In the future, it will not be necessary to speculate about which genes may be involved in the response to a medicine. Single nucleotide polymorphism (SNP) maps and associated technologies will enable the whole genome to be searched for genetic variations which are correlated with response. SNPs are positions in the genome at which two alternative bases can occur (Wang *et al.* 1998). These are the most common type of genetic variation in humans and occur at a frequency of approximately 1 per 1000 base pairs.

An ordered high-density SNP map of the human genome is being produced by a consortium of pharmaceutical companies, academic centres and a charitable trust. The SNPs in this map will be frequent (approximately 1 per 10,000 base pairs) and evenly distributed throughout the genome. Variations of several consecutive SNP markers that are close to-or within-a particular allele are likely to be present in patients with that allele, as they will be inherited together due to the phenomenon of linkage disequilibrium.

The distribution of SNPs in the SNP map will allow identification of SNP markers close to or within gene variants that can influence the patient's response to treatment (Evans & Relling, 1999). The genetic research being performed as part of a clinical trial will use the SNP map to identify those consecutive SNP markers in linkage disequilibrium that correlate with specific response to treatment. Identification of the actual genes and alleles involved would not be necessary. If abbreviated SNP profiles could be identified in phase II trials, it would be possible to select patients most likely to benefit from the treatment for the phase III trials. Such selection would allow the typically large phase III studies to be smaller and more efficient.

Pharmacogenetic research and ethical issues

The clinical studies to test the safety and efficacy of a new medicine must receive Institutional Review Board (IRB)/Ethics Committee

(EC) approval; both the protocol and consent form have to be approved. There must also be specific IRB/EC approval of the genetic research component and the consent form. Patient participation in the genetic research part of the study, as in the main study, is entirely voluntary. The patient has to provide explicit written informed consent before taking part in any genetic research that forms part of a clinical study on a potential new medicine. There are no study treatment implications: the patient's treatment is not affected by participation or otherwise in the genetic research.

Patient privacy and confidentiality are important to Glaxo Wellcome in all clinical activities including genetic research. The ethical procedures governing our genetic research were implemented after consultation with independent ethics experts. Throughout, patient privacy and confidentiality are paramount. The measures we have adopted incorporate the recommendations emerging from the United Nations 'Universal Declaration on the Human Genome and Human Rights' and the Human Genome Organization's 'Principled Conduct of Genetic Research', and the World Health Organization.

The genetic research procedures are reviewed frequently to take into account any changes in legislation, recommendations from the science and ethics communities and the needs of the clinical investigators and the patients. The patient's name and address are not transferred to the company; only a code number is used to identify the patient's medical data and genetic information. Because of the early stage of research, no individual results will be provided to anyone, including the patient, his or her family, employers or insurers, unless there is a legal requirement for disclosure. Once their significance has been confirmed, the group results will be published and shared with the medical community in a timely and responsible manner.

The move towards genetic research brings with it a need for educational material to help the investigators explain to patients about the research, the use of the sample taken, the risks and benefits and the procedures in place to protect privacy and confidentiality. The materials we have produced include, but are not limited to, brochures and videos for the patients and for the healthcare professionals.

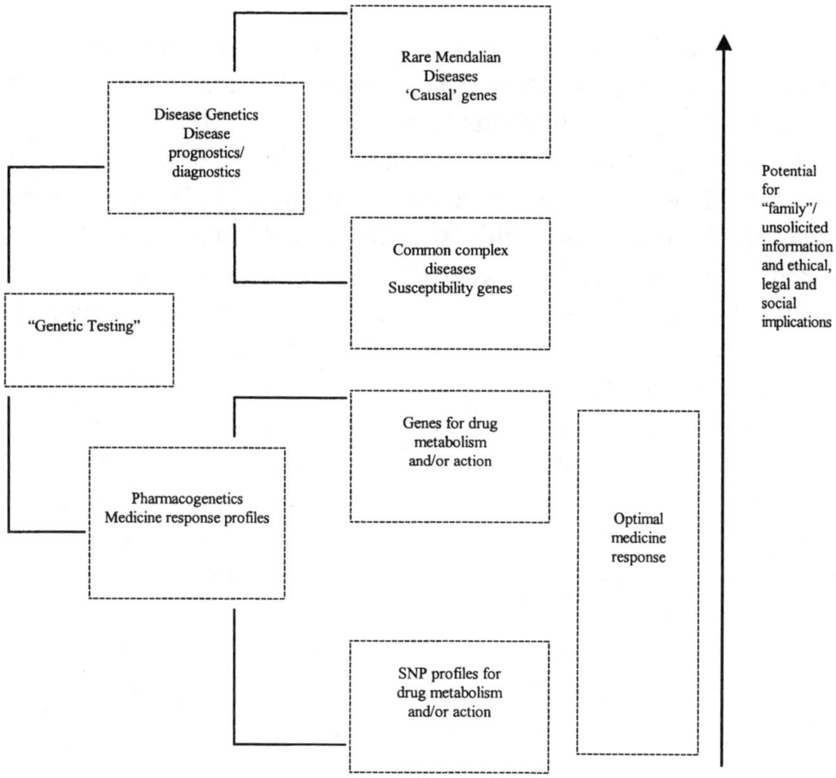

Figure 5.2 'Genetic testing' is used in a number of different settings

'Genetic testing' needs careful definition

One problem that has become apparent is that the phrase 'genetic testing' is used as a general term to cover different types of testing with quite distinct implications (Figure 5.2).

Under the broad heading 'disease genetics', there is testing for the rare Mendelian diseases (causal genes) and for the common complex diseases (susceptibility genes). The majority of the tests currently available and in development are specific for single gene (monogenic) disorders such as Huntington's disease. These tests detect the presence/absence of one or more mutations in a single gene. For such diseases, there is a strong causal relationship between the presence of a particular mutation and the development of the disease. The simple, Mendelian inheritance pattern of these diseases means that these tests have implications not only for the individual but also for family members, and these implications are exacerbated by the lack of effective interventions in almost all cases. The ethical, legal and social implications (ELSI) which arise include the potential for discrimination, the psychological impact on the patient and family, the implications for health care of patient and family and the need for pre- and post-test genetic counselling.

For common diseases, different alleles of susceptibility genes alter the risk of the disease developing. The example of the APOE gene in late-onset Alzheimer's disease has already been mentioned. Gene-specific tests for common diseases have the added complication that the presence of a disease susceptibility allele does not cause the disease but is a risk factor for the disease. The uncertainties generated (whether the disease will manifest and its severity) compound the complexity of issues such as patient discrimination and genetic counselling. However, it is clear that even under the heading of 'disease genetics', the different types of 'genetic test' have very different ELSI implications for patients and family members.

Pharmacogenetic profiles will predict the patient's response to a medicine. These profiles may take the form of gene-specific profiles and/or abbreviated SNP profiles that have been demonstrated to correlate with a response. The gene-specific profiles will determine the presence of gene variants that affect the mode of action and/or the metabolism of the medicine and that correlates with the response, whether that is a therapeutic effect or a side effect. In general, pharmacogentic medicine response profiles will be unlikely to provide additional information about the patient's disease or predict any other diseases. The various issues associated with this pharmacogenetic application of genetic science and technology are of a quite different magnitude from those associated with gene-specific tests for a disease.

Abbreviated SNP profiles will be produced by identification of regions of allele sharing in patients who receive the medicine and respond (either favourably or unfavourably) compared to those in patients who receive the medicine and do not respond. These medicine-specific profiles will not test for the presence or absence of specific gene variants and, for the most part, will not contain any genetic information other than that predicting how the patient will respond to treatment. Again, this type of pharmacogenetic application has quite different implications from gene-specific testing for disease: it can be likened to some clinical laboratory tests (such as blood typing and liver enzymes tests), which are already used to determine appropriate treatment.

In all cases of pharmacogenetic profiling, one of the major concerns with 'disease' genetic testing, i.e. the ability to predict disease without suitable treatment being available, is greatly attenuated. There will frequently be viable therapeutic options when a pharmacogenetic profile is used by a physician.

Thus, it is clear from this discussion that the ELSI issues associated with 'genetic testing' vary considerably, with predictive testing for Mendelian disease at one end of the spectrum and limited SNP profiling for pharmacogenetic use at the other. However, the debate on these ELSI issues is by and large driven by the Mendelian disease paradigm, and does not consider the full breadth of 'genetic testing' of the future. There is therefore the potential that perceptions and policy recommendations for genetic testing related to disease diagnosis could be inappropriately applied to pharmacogenetics. This could delay the delivery of the significant patient and health care provider benefits that will accrue from pharmacogenetics.

• Diagnosis (Disease by Symptoms)	→	Prognosis (Disease by Mechanisms)
• Guidelines & Formularies (Disease Uniformity)	→	Targeted Therapy (Disease Heterogeneity)
• Care Standards (Patient Uniformity)	→	Tailored Care (Patient Variability)
• Blockbusters (Universal Rx)	→	Mini Busters (Pharmacogenetics)

Figure 5.3 Impact of genetics on health care: The future

The anticipated impact of genetics on health care is summarized in Figure 5.3. We expect that there will be a move from diagnosis by symptoms to diagnosis by symptoms and mechanisms. Diseases will be seen to be heterogeneous and not uniform. Care standards will be tailored to the individual patient rather than standard approaches used for everyone. And perhaps there will be an end to the one-size-fits-all 'blockbuster' medicine-a medicine used to treat everyone-and emergence of the 'minibusters': medicines for subsets of patients based upon their pharmacogenetics.

Acknowledgements

I am indebted to all the members of the Clinical Genetics division at GlaxoWellcome for discussion, input etc.; however, special thanks for their contributions are due to Penelope Manasco, Elizabeth McPherson and Andrew Freeman.

References

Anderson, W.H., Fitzgerald, C.Q. and Manasco, P.K. (1999) Current and future applications of pharmacogenetics, *New Horizons*, 7, pp. 262-9.

Evans, W.E. and Relling, M.V. (1999) Pharmacogenomics: translating functional genomics into rational therapeutics, *Science*, 286, pp. 487-91.

Kuivenhoven, J.A., Jukema, J.W., Zwinderman, A.H., de Knijff, P., McPherson, R., Bruschke, A.V.G., Lie, K.I. and Kastelein, J.J. P. for the Regression Growth Evaluation Statin Study Group (1998) The role of a common variant of the cholesteryl ester transfer protein gene in the progression of coronary atherosclerosis, *New England Journal of Medicine*, 338, pp. 86-93.

Tan, S., Hall, I.P., Dewar, J., Dow, E. and Lipworth, B. (1997) Association between B2-adrenoceptor polymorphism and susceptibility to bronchodilator densensitization in moderately severe stable asthmatics, *The Lancet*, 350, pp. 995-9.

Wang, D.G., Fan, J.B., Siao, C.J., Berno, A., Young, P., Sapolsky, R. *et al.* (1998) Large-scale identification, mapping and genotyping of single-nucleotide polymorphisms in the human genome, *Science*, 280, pp. 1077-82.

6 The Use of Large Biological Sample Collections in Genetics Research: Issues for Public Policy

Paul Martin and Jane Kaye

Introduction

The concerns raised by the increasing use of large-scale biological sample collection for genetic research are profound. They touch on critical issues of human rights, personal identity, the future conduct of biomedical research, new forms of property rights and the proper relationship between academia and commerce. At the same time, and partly in response to these new scientific developments, public policy is in the process of transition.

This chapter has been written with a number of aims in mind. Firstly, it will attempt to describe briefly the scientific research strategies in the emerging field of functional genomics and the way in which investigators are using biological sample collections, geneological data and personal medical information to hunt for gene-disease associations. In particular, it will be shown that the use of sample collections cannot be easily separated from the use of medical records and data about family relationships.

The role of the biotechnology industry will also be briefly considered. Much ethical analysis of the issues raised by new genetic technologies is carried out in a social and economic vacuum, often without reference to the objectives of the powerful social actors who shape the field of research. In the case of biological sample collections, it is likely that industry will play a major role in the development of this field.

Following the introduction of the scientific and commercial background, a series of case studies of research involving the use of large biological sample collections and personal medical information will be presented. Much of the recent discussion of sample collections has been sparked by developments in Iceland and the activities of the biotechnology company, deCODE Genetics. This will be described and some examples of current or planned research in the UK will then be presented to highlight the key issues raised in the British context.

Many of the concerns being expressed about the potential misuse of the genetic information being generated by genomics research relate to the adequacy of the legal framework governing the conduct of investigation. The third section of the chapter will therefore present the current UK legal framework relating to the ownership of biological materials, patient confidentiality and consent. Important points about the adequacy of the existing legal framework in the light of new types of genetic research will be highlighted.

Finally, the key issues for further discussion and areas for future social science research which follow from both the case studies and the review of UK law will be summarized in the concluding section.

Genetic research involving

The post-genome sequencing research agenda

One of the landmarks in modern biology is due to be reached in the next few years when the sequencing of the entire human genome will be completed. However, the sequencing of the genome is only the start of a major programme of research, which is likely to occupy the biological sciences for decades to come. The next stage of investigation will be to understand exactly what the information coded in the human genome means and how this new knowledge might be used to improve health and health care.

Central to this task is the need to establish the function of the 100,000-140,000 genes contained in the 23 pairs of human chromosomes. At present, the biology of the great majority of the thousands of genes that have already been sequenced is unknown. Although the proteins which they code for can be identified, their role in both the normal workings of the body and in pathology is much harder to identify. Until the function of a particular gene and its role in pathology have been established, the raw sequence information is of little clinical value. This area of research has become known as functional genomics and is one of the most rapidly expanding areas of molecular biology.

The research required to link gene sequences (genotypes) to particular biological functions or diseases (phenotypes) is complex and involves a series of steps. Until recently, researchers have worked 'backwards', starting with the identification of an inherited pattern of a disease and then trying to find the genetic changes responsible for the condition. However, in most common diseases, the biological changes responsible for the pathology are not well characterized or it is difficult to demonstrate a simple pattern of familial inheritance. As a consequence, scientists have adopted a working hypothesis that some diseases, such as asthma, have a genetic component. Research then involves trying to make a *correlation* between having the disease and carrying a particular gene sequence or genetic change.

This type of research is made more difficult by the fact that only a subset of some common diseases may have a clear genetic component to them. The challenge for research is therefore to try to identify which sub-groups to study in large populations of people suffering from a given disease. In cases of, say, familial cancer, this is fairly straightforward. However, in other diseases, where there is no clear pattern of inheritance, this is a demanding task, as the starting assumption may be incorrect.

Identifying diseases which show patterns of inheritance

The starting point for most research aimed at identifying diseases which have an inherited component are studies of families or small groups where there is evidence of a higher than average incidence of a particular condition. Family studies of this sort have been routinely undertaken by clinical geneticists in the UK, and include research on rare monogenic disorders, familial cancers and familial forms of other common diseases.

However, relatively little work has been done on the inheritance of more common diseases. In order to investigate whether some diseases previously thought of as being acquired have inherited forms, researchers have recently started to hunt for small groups or populations that appear to suffer from a high rate of a given disease. Such communities may often be geographically isolated, resulting in a level of inbreeding and genetic homogeneity which makes genetic studies easier.

Through the use of family, group and genealogical studies, it has become possible to identify sub-populations of relatively common diseases that have a clear genetic component. Many of these conditions had not previously been thought of as 'genetic' in any sense.

Identifying the gene sequences associated with inherited diseases – linkage studies

Where a clear pattern of inheritance of a disease can be established, powerful genetic mapping techniques can then be used to identify the genetic changes (mutations, deletions, polymorphisms etc.) which are associated with the pathology.

Instead of studying the total human DNA sequence, analysis is simplified by examining a relatively small series of short DNA marker sequences spread evenly across the entire genome: a process known as genotyping. The most common markers are called single nucleotide polymorphisms (SNPs) – see below. The pattern of inheritance of these markers is then related to the pattern of inheritance of the disease within the families studied. Using this approach, it is possible to identify particular small regions of chromosomes which contain the genetic change associated with the condition. Automated gene sequencing and the use of gene sequence databases can then be used to identify the possible genetic change involved.

Making a correlation between a disease and a specific genetic change in cases where no pattern of inheritance is obvious – association studies

The basic principle behind genetic association studies is the statistical correlation between specific DNA sequences and particular diseases in cases where there is no clear pattern of inheritance. As a consequence, fairly large groups of people suffering from the disease have to be studied, as only a sub-population of the pathology may involve a strong genetic component or, alternatively, the genetic influence could affect many people, but may not be very marked.

Instead of trying to make disease-gene associations using raw sequence data, researchers are starting to use SNPs. It has been found that human populations are to some extent genetically heterogeneous, i.e. the exact sequence of a particular gene varies within a population. The variation is generally limited to a relatively small number of single base pair changes (SNPs) which are stable and inherited across generations. Many of these SNPs are not harmful, but some appear to be deleterious and may be involved in causing specific diseases or adverse drug reactions.

The overall strategy guiding association genetics involves collecting samples from patients with a specific disease and then genotyping their DNA using large arrays of SNP markers. The hope is that specific SNPs will be found to correlate closely with the disease being investigated. This analysis is also carried out on individuals not suffering from the disease, to provide a control group.

Studying the interaction between genes and environment – genetic epidemiology

In recent years, new fields such as 'genetic epidemiology' and 'molecular epidemiology' have emerged, which seek systematically to apply the traditional methods of epidemiology to the study of environment – gene interactions and their role in pathology. As greater knowledge of genes associated with diseases and human genetic polymorphisms is gained, it will be increasingly possible to analyse the role which genetic risk factors and specific environmental hazards play in the cause of common conditions such as cancer or heart disease. It is clear that this type of research will depend on very large population – based sample collections and access to detailed patient information (i.e. medical records).

The use of human biological sample collections by the biotechnology and pharmaceutical industry

The majority of companies working in this area are primarily concerned with generating and selling information about the relationship between specific genetic sequences and particular diseases, rather than developing drugs themselves. However, some firms are planning to develop diagnostic tests based on this data, a number are offering contract genotyping services, and others are looking to develop drugs in partnership with large pharmaceutical companies.

Table 6.1 gives information about the strategies of some of the leading European biotechnology firms working with large sample collections. It illustrates the mixture of strategies being adopted, with the main focus on the hunting of genes associated with common diseases and studies of pharmacogenetics. The most popular disease targets include cancer, cardiovascular diseases, depression, schizophrenia and osteoporosis.

It should be highlighted that private sector activities depend heavily on both public-funded research and widespread public participation. It is therefore difficult to disentangle public and private research, as researchers from both sectors are often involved in supporting the same project. Very close academic-industry links are a general feature of research in human genetics. Whilst this enables effective technology transfer, it also gives rise to concerns about academic conflicts of interest and commercial monopoly.

Table 6.1 Strategies of selected European firms working with large sample collections

Firm	Business focus/disease areas	Scientific strategy	Samples collected	Personal medical information used
deCODE Genetics (Iceland)	Identification of genes linked to common diseases. Pharmacogenetics. Sale of data and development of diagnostics. 35 disease targets.	Linkage and association studies (SNPs).	10,000 samples taken from Icelanders with specific diseases.	Use of limited data about disease at present. Planned use of comprehensive database of all population's medical records.
Eurona Medical (Sweden)	Pharmacogenetics. Sale of data, related diagnostics and predictive tests. Hypertension, cancer, depression and schizophrenia.	Location of genetic variants associated with ADRs.	Access to over 3 million samples and related medical records.	Data about therapy, outcomes and adverse reactions.
Gemini Research (UK)	Identification of genes linked to common diseases. Sale of data and development of diagnostics. CV disorders, obesity, osteoporosis.	Association studies (SNPs).	Collection of samples from several thousand non-identical twins.	Very detailed clinical information (over 900 data points)–much collected during research.
Genset (France)	Identification of genes linked to common diseases. Pharmacogenetics. Sale of data and development of diagnostics. Cancer, schizophrenia, depression, Alzheimer's, obesity, CV disease, osteoporosis.	Linkage and association studies (SNPs). Location of genetic variants associated with ADRs.	Access to collections in USA, Israell, Argentina, France and Germany. Irish collaboration using >10,000 samples from CV disease patients.	Varies according to study, but would involve access to full patient records in several cases.
Oxagen (UK)	Identification of genes linked to common diseases. Pharmacogenetics. Sale of data and development of diagnostics. Osteoporosis, endemetriosis, inflammatory bowel disease, coronary artery diseases.	Linkage and association studies (SNPs).	Families with high incidence of disease. Studies planned with samples from up to 10,000 individuals.	Use of limited data about disease at present. Planned development of large database of genotypes and outcomes.

Case studies of the use of biological sample collections and personal medical information in human genetic research

A series of four case studies will be presented in the following sections to provide concrete examples of the type of research being undertaken and the ethical, social and legal issues raised by these developments. The first example will the work of deCODE Genetics in Iceland and the creation of the Icelandic Health Sector Database. This will be followed by brief descriptions of examples of current or planned research in the UK.

DeCODE Genetics and the creation of the Icelandic Health Sector Database

Much of the recent international discussion of the issues raised by the use of biological sample collections has been stimulated by developments in Iceland. In particular, a proposal for an electronic database containing detailed information from the entire population's medical records has been championed by a biotechnology company, deCODE Genetics. This has aroused widespread fears about the potential abuse of human genetic research.

DeCODE Genetics is a private company founded in 1996. The firm operates out of Reykjavik and employs nearly 300 staff. It was created specifically as a 'population-based genomics company conducting research on the causes of common diseases'. In particular, it aims to exploit unique features of Iceland's population, as well as the country's extensive genealogical records and high quality health care system. Most Icelanders are descended from a very small number of individuals who settled in the country in the 9th century. This has resulted in a high level of genetic homogeneity and greatly simplifies the technical problem of trying to identify these disease-related gene variants.

In order to carry out the genetic analysis of common diseases, the company has established two core technologies:

- *A computerised genealogical database* – containing records of 600,000 individuals (living and dead) and their family relationships.
- *High-throughput genotyping* – the ability to process and scan DNA samples using large numbers of genetic markers.

The first phase of the firm's research strategy, which has been underway for several years, has been based on collaborating with local doctors to collect DNA samples from people suffering from particular diseases. The genealogical database is then used to cluster these patients into large extended families, thus allowing genetic linkage analysis to be undertaken using high throughput genotyping. As of September 1999, the company had collected samples from over 10,000 people with full written consent.

By 1999, deCODE had already established research programmes on the genetics of 35 common diseases, including cancer, myocardial infarction, heart disease, multiple sclerosis, diabetes, osteroarthritis, Alzheimer's disease and schizophrenia. Twelve of these programmes are in collaboration with the pharmaceutical company Hoffmann-La Roche, who is paying $200 million over 5 years for access to the findings. Already the company has successfully used this strategy to identify a gene involved in an inherited form of endometriosis.

The second phase of the firm's research strategy will involve the construction of an electronic health information database, the Icelandic Health Sector Database (IHD), which when built will contain the encrypted medical records of almost the entire population. Work has not yet started on this, and it is not likely to be complete for several years.

The right to construct and operate the IHD will be licensed from the Icelandic government, which has passed specific legislation on this matter, and it will be financed entirely by the company. In December 1998, the Icelandic government passed the Act on a Health Sector Database 1998.[1] In return, deCODE will have the sole right to exploit this resource commercially for a period of 12 years.

The proposal for the Icelandic Health Sector Database

The proposal for the IHD contained the following elements:

- The database would hold personal medical information on all citizens and would be based on their medical records.
- The information would be held in an anonymous form that would prevent the identification of individuals.
- The right to build and operate the database would be licensed by the government for a fixed period.
- Its operation would be carefully regulated through the licensing agreement and a series of government agencies.
- The database would not be linked to other external and unrelated databases but would comprise the three related datasets (the genealogical database, medical records and genotyping data).

According to deCODE, the IHD will contain information on:

- 'Longitudinal disease progression.
- Treatment and treatment response.
- Direct and indirect cost of treatment and cost effectiveness'.[2]

This main emphasis would be on health-related data which can be coded and would include medical information, health resource use, genealogical information and genotype data. Diagnoses, test results, data on forms of treatment, side effects, response to treatment, duration of therapy, and place of treatment would all be entered, allowing costs to be calculated. However, as of October 1999, many details of the database were still unresolved including, exactly what information it will contain and how it will be used.

The public debate on a Health Sector Database

The proposal for the Health Sector Database has been highly controversial both in Iceland and internationally. Over the past 18 months, there has been considerable debate in the Icelandic Parliament (the Alpingi), in the electronic media and press concerning its creation. Icelandic groups opposed to the plan, such as Mannvernd and the Medical Association have galvanized international criticism of the proposal and the world's media have led with headlines such as the 'Selling the family secrets'[3] and 'A human population for sale'.[4]

The database has been controversial for a number of reasons, including:

- The legislation was drafted quickly and without widespread community consultation, and there were rumours that deals had been done behind closed doors.
- The original proposal did not allow people the opportunity to consent to the use of their personal information and did not allow them to opt out of the database. This has since been changed, as people now have the opportunity to opt out within a fixed period.
- The original proposal was to encode the medical records rather than make them anonymous, which meant they could more easily be traced back to individuals. In a small country like Iceland this would have major implications for privacy.
- There were fears about the security of such a database.
- The Icelandic scientific community was worried about scientific freedom and the effect on research of putting access to medical records in the hands of a private company.
- The fact that the personal information of a whole population was to be collated by a private company, with no clear statements as to how the database would be used, was alarming.
- There was scepticism about the promised benefits to Iceland and concern that a single company was being given a potentially lucrative monopoly.

Consent for inclusion in the database

Despite continuing criticism, the Icelandic government has chosen not to seek the consent of individuals before including their medical records on the database. It will instead rely on an 'opt-out' process. While there has been much criticism of the decision not to seek consent, this has been an accepted research practice for epidemiological research in both the UK and internationally.

The general view in the UK is that personal information in such studies should be anonymous, and if the research does not harm the individual and a research ethics committee has given approval, then consent is not required.

The design of the database – making the information anonymous

Much concern has focused on how to make the database anonymous to ensure privacy, while at the same time allowing new data to be added to existing records on the database. In order to ensure that the data is anonymous, information will pass through three layers of coding.

The Data Protection Commission will be responsible for overseeing the linking of the Health Sector Database to other datasets. DeCODE want to link the database with the Icelandic genealogies that are in the public domain, and existing sample collections that the company has collected. Exactly how this will be done to prevent the joining of personal identifiers is still to be worked out. The concern is that in a small country such as Iceland individuals are more recognizable than in larger communities and that the joining of different databases would enable individuals to be identified. It is also recognized that computer systems are fallible, and that security cannot always be guaranteed. One measure designed to minimize this risk is that it will not be possible to extract information from the database of groups smaller than 10 individuals.

The use of the database and access by third parties

The exact procedures that will be put in place for access to the database are still not finalized. They will be the subject of regulations that are in the process of being drawn up. The Healthcare Sector Database Act requires that all research questions put to the database must be approved by a Scientific Ethics Committee, which applies equally to deCODE Genetics as well as other parties.

While deCODE Genetics will have the right to charge third parties for use, the company will not have the right to exclude, and it may find that its competitors would have to be given access to the database.[5] In addition, The Ministry of Health and the Director-General of Health will always be entitled to statistical data from the Health Sector Database free of charge. However, it is not clear whether researchers based in Iceland would have to pay for access to the database or whether deCODE will allow them access free of charge but subject to ethical approval.

UK genetics research using large biological sample collections

This section will briefly describe three examples of existing or planned UK research projects involving the use of large biological sample collections. At present, there is no project in the UK which is equivalent to the Icelandic example, but the proposed creation of the UK Population Biomedical Collection by the Wellcome Trust and the MRC has many similarities. The examples described below have been chosen to illustrate the type of research being undertaken and the issues raised, but are not intended to provide a comprehensive picture of research in the UK.

Initiatives under consideration by the MRC and Wellcome Trust

In 1998, the MRC received increased funding as a result of the government's Comprehensive Spending Review. In particular, it included in its bid a proposal to support national DNA collections as part of its Post-Genome Challenge. Several ideas were discussed, including the creation of a very large population study. The additional funding including £12 million earmarked for the creation of DNA collections.[6] As of the end of 1999, the Council was still planning how to support this type of research and was working closely with other funders of biomedical research, including the Wellcome Trust.

As of December 1999 no detailed information about the MRC's plans were in the public domain and many of the issues relating to the creation of large-scale sample collections were still being discussed. However, two main possibilities (which are not mutually exclusive), were under consideration. The first is the funding of a series of regional DNA banks, which could then be used as a resource by the biomedical research community as a whole. Samples would come from a range of studies funded on a project-by-project basis, and might also include samples from existing collections if appropriate consent had been obtained from donors. This would probably occur through support for 'private' collections by consortia or individual scientists, who would then be obliged to split samples and place part in the regional DNA bank. The banks would offer a genotyping service and access to data, but not to the samples themselves. No plans exist to link these banks in a systematic manner to medical information from NHS sources, but associated data characterizing the samples will include medical information.

The second option involves the creation of a single very large new resource, the UK Population Biomedical Collection, in collaboration with other funders.[7] This proposal was first discussed at an expert workshop in May 1999 organized jointly by the Wellcome Trust and the MRC. It would be focused on genetic epidemiology and the collection would enable prospective studies of genetic and environmental risk factors in diseases of later life. As a consequence of the need to analyse both genetic and environmental factors simultaneously, and the interaction between them, the proposed collection would be very large, containing samples from up to 500,000 individuals.[8] The prospective nature of the study would also span many years and require the ongoing collection of data from research subjects.

If such a large population collection were created, it would be a collaborative national effort. Genotyping would be done in centralized facilities and investigators would only have access to data, not to the samples themselves. All proposals for research would be peer reviewed. Companies would be able to access the data from the collection, but only on a non-exclusive basis. However, the issues surrounding getting access to personal medical information and the prospective nature of the research have not been resolved, and no final decision about the creation of this resource will be taken until Summer 2000. In addition, no details of the oversight of this proposal have been made public.

Oxagen

Oxagen is a private biotechnology company based in Oxford established to investigate 'fundamental insights into the molecular biology of common human disease'. It was founded in 1997 as a spin-out from the Wellcome Trust Centre for Human Genetics at the University of Oxford. It also has very close links with the Nuffield Department of Clinical Medicine, the largest group of clinical researchers in the UK.

In collaboration with leading clinical research groups Oxagen is creating large, well-characterized sample collections from families with 'predispositions to specific diseases'. These collections are then used in conjunction with family pedigree information and clinical data to perform linkage analysis using high-throughput genotyping.

The company is interested in three broad areas: women's health, coronary artery disease, and inflammation and autoimmune diseases. Its programmes are spread throughout Europe and the US, and involve 31 collaborators in 22 centres. In women's health, Oxagen has signed a 5-year collaboration with six European bone research groups to identify 'osteoporosis-related genes', and 3000 osteoporosis sufferers and family members are being recruited for investigation. In its coronary artery disease programme the company plans to collect samples from over 10,000 affected and unaffected family members.

Its research is carried out in the following manner. All of its clinical research is done in collaboration with academic investigators, many of whom have already been working with family sample collections for some time. The clinicians undertake all direct contact with participants and Oxagen never sees patients. Full consent is obtained, and research subjects are also told that a commercial company is involved, and they are asked to disclaim any rights to future financial gain arising from the research. However, Oxagen stresses that it does not ask them to disclaim ownership of their sample and medical records. The ownership of samples therefore remains in the public domain. At any time, patients can withdraw samples from the study by asking the doctors involved. Furthermore, the company cannot link samples to the names of participants, as only coded materials and information is passed on by the doctors. In effect, the clinical collaborators provide a Chinese wall, which ensures confidentiality.

Consent is only given for a specific application/study in a given disease area and the company can only use the sample in a tightly defined manner. The use of the complete collection is also governed by a research steering committee, which has a majority of academic members. The only stipulation is that Oxagen has an exclusive license to commercially exploit the collection for a fixed period of time (usually 3-5 years). It is therefore asking for a 'commercial head start' and not a monopoly over the use of the collection. Oxagen funds the process of collection and has agreed to share milestone and royalty payments from deals with large partners.

Every proposal for research has to receive prior approval from a research ethics committee (usually an MREC). In addition, patients do not get any feedback of the results of the research.

Oxagen only makes use of fairly limited clinical information in its research. Permission to access medical records is included in the consent form, in line with standard Good Clinical Practice (GCP). No medical records are received by the company and the only information used by Oxagen is contained in a questionnaire completed by the clinical collaborators. The form is coded, so that no personal identifiers are present and the same is true of the samples and pedigree data. Information is therefore only stored in a coded form by the firm.

The UK legal and policy framework governing genetic research using large biological sample collections and personal medical information

There is no one piece of legislation that oversees the establishment and use of DNA Banks in the UK. Instead regulation is largely dependent on the guidelines of the medical professional bodies as the current law offers little protection for the rights of individuals to control the secondary use of biological samples or anonymous personal information. The result of this is that individuals may not know that their biological samples are being used in research or that their personal information is being used for a research purpose that they may not have consented to. This is of some concern as databases for genomic research are increasingly being used by industry. Private companies and their employees are not necessarily bound by the guidelines of the medical professional bodies, and in this area there is very little law to regulate their activities. The purpose of this part of the chapter is to outline the law that covers the use of biological material and personal information in medical research and to discuss some of the policy issues that are of significance.

The legal framework that governs medical research in the UK

In the UK, medical research is controlled through a system of self-regulation by the professional medical bodies with recourse, if necessary, to the courts. There is no legislation in the UK that regulates research on human beings, although there is legislation that covers research on animals.[9] Despite being able to appeal to the courts, there have been few cases that have been directly related to the conduct of research. This means that the law applying to medical research is largely dependent on the common law principles of consent and confidentiality developed for treatment, and the European Directives that have been implemented into UK law. The European Convention on Biomedicine and Human Rights,[10] which deals specifically with biomedical research has still not been signed by the UK[11] and so, while it is authoritative, it cannot as yet be considered to be a part of UK law.

It is the guidelines issued by institutions such as the Royal College of Physicians of London, the Medical Research Council, and the Department of Health that set the standards that determine how medical research is carried out. These guidelines have a quasi-legal status and, if a case ever reached the courts, they would be considered when determining the lawfulness of research practice. These guidelines attempt to plug the gaps where there is no common law or legislation to provide direction. This means that the guidelines can have higher standards than the law and greater consistency in their approach to specific issues. The benefits of having a system of regulation based on guidelines is that they can be adapted to meet changing circumstances. However, the drawback of such a system is that it will not apply to those who may be working outside the medical framework in private companies. In addition, different bodies may have different guidelines for the same issues, so people working alongside each other could be governed by different principles. An example of this are the guidelines that have been drawn up over the past year by the Medical Research Council,[12] the Royal College of Physicians of London,[13] and the Royal College of Pathologists[14] regarding the use of biological samples in research.

Medical research is regulated by Local Research Ethics Committees (LRECs) and Multi-centred Research Ethics Committees (MRECs), which draw on these guidelines to monitor how research proposals are carried out. These bodies are comprised of people from different backgrounds and make decisions that can reflect local concerns and therefore may vary across the country. The decisions of the research ethics committees do not have legal standing, as their role is to determine ethical questions within the parameters of the guidelines and the law. Furthermore, the committees do not have any statutory powers of investigation by which they can enforce their decisions. Any breach of ethical standards, if detected, would be referred to a disciplinary panel of the appropriate professional body or the judicial system.

Currently, much of the research done in the private sector is regulated in much the same way as public research. However, this is done on a purely voluntarily basis, as there is no legislation or statutory oversight mechanisms governing how private research is carried out, although there is the Data Protection Act 1998 on the use of personal data. Similarly, there are people working within medical research who are not bound by any professional codes of conduct or guidelines. Private research has the added factor of having to maintain commercial trade secrets that could make effective monitoring and oversight difficulty. For example, relatively little is known about the increasingly common practice within drug company-sponsored clinical trials of routinely collecting and storing samples for pharmacogenetics research.

Biological samples

Biological tissue samples are parts of the human body that are removed by 'the aspiration of bodily fluids (for example, blood) through a needle, by the scraping of cells from a surface (for example, skin or cervix), surgical removal (such as organs or biopsies), or collection by non-invasive procedure (e.g. semen)'.[15] A sample that becomes part of a tissue bank might be left over from an operation, collected as part of an autopsy, have been used for diagnosis, donated for research, or become part of an archive. The way in which a sample is obtained and the consent that accompanies it can have implications for its subsequent use in research. Different law applies depending upon whether a sample is derived from a living person or removed from a cadaver.

The removal of biological samples from the individual

(a) Removal from the living

The law in this area is based on common law principles developed for treatment rather than research. The UK legislation that covers removal of tissue from the living is the Human Fertilization and Embryology Act 1990, which regulates the use of embryos and gametes in research, and the Human Organ Transplant Act 1989, which covers the removal of organs. All other tissue removal from the living is covered by the common law doctrine of consent that seeks to protect individual autonomy and bodily integrity. The common law is based on the principle of 'broad consent', which is a lower standard of consent than that requirements of legislation or the informed consent laid down in the Declaration of Helsinki. This standard allows a doctor to remove a number of samples from an individual at the same time, without consent for all samples, and to use a sample for a different purpose than that for which it was collected.

The principle of broad consent means that the removal of tissue will be lawful and not constitute a battery if the patient understands in broad terms the nature of the procedure to which he/she has agreed.[16] For example, a doctor is able to remove a sample from an individual for diagnosis and another one for research, and this would not be considered a breach of consent.

This principle of broad consent means that a biological sample removed from the body with consent could be used for a secondary purpose and that this would not be unlawful. For example, if consent is given for the removal of blood and then a genetic test is carried out on the blood sample without further consent, this would not be unlawful if the physical removal of the blood through the syringe was done with consent. In the USA, the courts have applied this reasoning in situations where the secondary testing has been for the individual's benefit. While this is not unlawful in the UK, it is contrary to the guidelines of the General Medical Council, which requires that procedures should be carried out with informed consent. The General Medical Council has recently reprimanded a GP for testing five patients who presented with different conditions for HIV without their consent.[17]

The common law does not require that a doctor provide all the information necessary for an individual to make a decision. Instead, a doctor has a duty of care to disclose to a patient the 'material risks' associated with a medical procedure prior to acting, to ensure that proper consent has been given for those acts. In the UK, the doctor should inform the patient of that which doctors as a profession think it appropriate for the patient to know.[18] In contrast, the doctrine of informed consent focuses on the individual and the quality of the information he/she needs to know about the potential benefits and risks of a procedure before he/she can make an informed decision. This means that, in the UK, in the case of medical treatment a doctor has the discretion to withhold information from a patient about a physical intervention. Under the common law, a doctor would not be required to inform the patient that a biological sample could in the future be used for research purposes, particularly if there was no risk of misadventure or injury to the individual as a result of that research.

(b) Removal from the dead

The Human Tissue Act 1961 regulates the use of human tissue for the purposes of medical education or research. In order for tissue to be removed from an individual, the Act stipulates that there must be express consent either in writing or orally from the individual during their previous illness that has been witnessed by two people. However, if the 'person lawfully in possession of the body' believes after making 'such reasonable inquiry as maybe practicable' that the deceased has made no express objection to the use of their human tissue and the relatives of the individual do not object, then the tissue can be removed from the body for research purposes.[19] The ambiguity of this requirement has led to uncertainty about whether consent for removal of tissue samples for research has actually been obtained. A public inquiry has begun at the Alder Hey Hospital in Liverpool, where there has been concern as to whether the removal of organs from dead children had been done with the appropriate consent of the parents.[20] The Human Tissue Act offers no sanctions for tissue that is removed unlawfully and contrary to the Act.

The other pieces of legislation that allow the lawful removal of human tissue are the Anatomy Act 1984, which allows removal of tissue for the purposes of 'anatomical examination',[21] and the Coroners Act 1988, which allows post-mortem examinations to be carried out.

Control of biological samples in tissue banks

The law of the UK offers very little or no guidance in this area, as there is no legislation to cover the use of biological samples in tissue bank collections. The common law position is that there are no property rights in the body, so no one can own the body or its parts not even the person from which the parts are derived. The courts have found some exceptions to this rule, usually in cases of theft. It has been found that there are property rights in stolen hair,[22] urine[23] and blood[24] samples. Relatives who need to dispose of a body also have a property right in the corpse. As an extension of this line of theft cases, the recent case of *R v. Kelly*,[25] established that, where body parts have acquired different attributes "by virtue of the application of skill, such as dissection or preservation techniques, for exhibition or teaching purposes", then the body parts became property and were able to be stolen.

This decision has implications for the use of DNA samples derived from dead bodies, as it suggests that, if there has been the application of skill and work in isolating the DNA, then the person or institution that does so gains a property right over the DNA sample. This is a more comprehensive right than the personal right of the individual to bodily integrity that is protected by the common law doctrine of consent. A property right entitles the owner to use or exploit the thing, to protect the thing against unauthorised use, and to allow transfer by gift or sale. This means, once the DNA has been isolated from a sample, it can be used without permission for research by an institution or a company. This in effect gives greater rights over the sample to the person or institution that has isolated the DNA, than the individual from which it was derived. It is not clear whether this case could also be applied to tissue from living individuals and the context of DNA banking. If it did, this would raise a number of ethical concerns about the protection of the interests of individuals, and whether it is appropriate that individuals have no rights over the use and control of excised tissue samples. This is of special concern because of the nature of the genetic information that can be derived from tissue samples. However, the court in *R v. Kelly* was not prepared to challenge the no property rule in the body, preferring to leave this to Parliament and the introduction of legislation.

It is not clear at what point the Data Protection Act 1998 provisions would apply to biological samples and the information that was derived from them. The definition of data under the new Act is considerably widened, as it now includes identifiable personal data intended to go onto a computer, as well as paper filing systems. It is unclear whether biological samples in their physical form would come under the Act; however, they may come under the Act if they were scanned onto a computer. If this were the case, samples for tissue bank collections would have to be acquired with explicit consent for the processing, which could have implications for existing collections derived from surgical waste or diagnosis where consent for research has not necessarily been obtained.

Personal information

It has been argued that the common law doctrine of confidentiality and the Data Protection Act 1998 provides a seamless protection for the individual against unauthorized use of personal information.[26] The law does not protect individual interests in the use of anonymous information, as it is seen that this does no harm to an individual. However, just because personal information is removed of personal identifiers, it does not mean that the information loses its quality of confidence or that an individual no longer wishes to specify what it can be used for. In medical research, the fact that there is no harm done to the individual because they cannot be identified has been the basis for not seeking consent for some types of research such as epidemiology.

(a) Confidentiality in the common law

The common law duty of confidence protects the individual against the unauthorized use of his/her personal information which is not public property or knowledge. The doctrine requires that such information, when disclosed in a confidential manner, must not be given to a third party without the consent of the person concerned, or the person who is authorized to act on the patient's behalf. However, there are exceptions to this basic principle.

There will be no breach of a confidence if explicit consent is given, either in writing or verbally, to the disclosure of the information. Another exception is if the disclosure can be justified in the public interest, because someone or the public at large may be put in danger by the patient.[27] Statute may also require the disclosure of confidential information, e.g. Abortion Act 1967 and Public Health (Control of Diseases) Act 1984. The recent case of *R v. Dept. of Health ex parte Source Informatics*[28] reiterated the principle that, if information is anonymous, there can be no breach of confidence if it is then passed on to a third party.

(b) Professional codes of conduct

There are professional codes of conduct that stipulate the duties of health professionals regarding confidentiality. The Department of Health has introduced the system of Caldicott Guardians to improve the standards of confidentiality in the National Health Service.[29] To breach these codes may result in disciplinary actions against individuals. However, different health professionals may be governed by different guidelines, according to the association they belong to. While guidelines will cover most health professionals, they do not cover all the staff who may be involved in the research project.

(c) Legislation

The Data Protection Act 1998 came into force on 1 March 2000, with various sections coming into effect over the next 2 years. The Act does not apply to personal data relating to a deceased person or anonymous data. However, if anonymous data are put with personal data that ultimately permit identification of a living person, then it will be covered by the Act. Unlike the previous Act of the same name, the 1998 Act is not restricted to computer records but also includes paper filing systems such as medical records.

The Act requires that consent must be obtained for all processing of personal data and that the processing shall be done fairly and lawfully. The individual must be told the identity of the organization processing the information, the purposes for which it is being used and to whom it will be passed, and this must be done within a time that is practicable. For the processing of sensitive data relating to a person's physical or mental health, sexual life, racial or ethnic origin, religious beliefs, or (alleged) crimes, the standard if higher and explicit consent must be obtained for its processing and use.

Under the exemption, the processing of sensitive data must not be "measures or decisions targeted at particular individuals and must not cause substantial distress or damage to a data subject".[30] The exemption allows personal data to be used for secondary research purposes without informing the individual. The personal data can also be held indefinitely and the individual has no right of access to his/her personal data in order to rectify any inaccurate information. Any research results that are published must not identify an individual.

Under Article 8 the Human Rights Act 1998, there is a right to respect for private and family life. This has implications for the protection of privacy in the UK, but it still remains to be seen how the courts will develop this concept.

Overview of the ethical, social and legal issues raised in the UK

The following section provides a summary of some of the key questions which are raised by the creation of large-scale biological sample collections for genetic research.

Conceptual issues raised by genetic associations with common diseases

The claims of recent work in genetics challenges previous assumptions about the underlying causes of many common diseases. Cancer is a good example of this. Twenty years ago, only a handful of tumours were thought to be caused by inherited factors but, in the last decade, a genetic model of the disease has started to dominate biomedical thinking. This new explanation is not based simply on the inheritance of cancer-causing genes. Instead, the description of the underlying pathology is now couched in molecular terms (patterns of gene expression, activation of oncogenes etc.). The role of environmental factors is still included in the model, but the emphasis has shifted to understanding how therapeutic interventions can be made at the molecular/genetic level. This 'genetification' of pathology is occurring in the explanation of many other common conditions that had previously been seen as acquired as a result of environmental hazards and social factors. Far greater attention is now being paid to genetic predispositions and the inheritance of disease-'causing' genes. This shift is likely to have profound long-term implications for our understanding of health and illness, and the conduct of medicine.

What are the implications of the shift towards genetic explanations of common diseases for clinical practice and our understanding of health and illness? What are the philosophical and theoretical issues raised by the shift towards molecular pathology? How are new diseases categories 'socially constructed?'

The use of large biological sample collections by academia and industry

This chapter has sketched out the main ways in which biological sample collections are likely to be used in genetic research in the immediate future. Their use in medical research in general and genetic research in particular is pervasive. However, very little is known about the scale of the collection and storage of biological samples in the UK.

- Where are large collections of biological samples held in the UK? How are they being used in research and for other purposes?
- What is the scale of sample collections being created by industry for use in genetic research? Should there be a system of monitoring such large-scale private collections?

- What other types of genetic research outside the field of functional genomics and genetic epidemiology are making use of large sample collections?

The creation and use of biological sample collections in genetic research

Consent, and the creation and use of collections. The principle of consent is central to the process of creating and using biological samples in medical research. However, there is some debate as to whether the standard of fully informed consent can be met in ever research situation. There are several reasons for this. Firstly, the complexity of genetic research makes it difficult for participants fully to understand the nature of the study they are involved in. Secondly, at the time that new collections are created, it is difficult to foresee all the potential research applications that the collection may be used for. Finally, in the case of existing collections, it may be impractical to gain consent for new research uses from the donors of the samples.

Other important issues are raised by the use of anonymous samples, which have previously been seen as relatively unproblematical. Given the nature of genetic information, it may prove impossible to ensure that biological samples can be truly anonymous. This underlines the point that guarantees of confidentiality can never be absolute and raises the question of the degree to which it is reasonable to take steps to protect confidentiality.

Within Western medicine, consent has historically been seen as purely a matter for the individual receiving treatment or participating in research. However, genetic information about an individual is also shared with other family members and may have implications for communities. This potentially sets the rights of the individual against the interests of their family. Furthermore, if a family or specific community is to be the subject of research, then there is a case for saying that consent may be required at a group level.

- Is it possible to have informed consent given the complexity of genetic research?

- Is it ethically acceptable that broad consent for the use of tissue samples is the standard, which would mean that individuals would not need to be informed of every new type of research conducted on the collection?
- Does the nature of consent have to be reconsidered in the light of genetic research?
- For tissue that is gained as surgical waste or has been archived, is it ethical that these collections can be seen as abandoned by the individuals and therefore future research does not require consent?
- Does the nature of genetic research, with its implications for other relatives, mean that consent should be considered as not just being relevant to the individual?
- Should this principle of community consent be recognized in the UK?

Privacy and data protection Some of the most important concerns about the creation of the Icelandic database have centred on issues of privacy and data protection. In particular, there are doubts that the highly sensitive information about the genotypes of individuals can be kept fully confidential. The use of coded samples in the UK raises similar questions, as in some situations it is not difficult to relate a sample to an individual, and it is only internal research policies and practices which prevent this from happening routinely.

Another set of concerns surrounds who has access to sample collections and databases of genetic information. DeCODE plans to sell subscriptions to its database to pharmaceutical and insurance companies. In the UK, access by third parties to databases are determined by private agreements and the requirements of the Data Protection Act 1998. The conditions of each agreement regarding database access will vary and depend upon negotiations between parties. Oversight of these private contracts is only done by a court when things go wrong.

- Is it possible to integrate large databases of anonymous genetic, medical and family information in such a way that confidentiality can be maintained?

- Is the use of coded samples adequate to protect confidentiality?
- Will it be possible to attribute unidentified samples in the future?
- Should there be a right to genetic privacy?
- How should access to genetic databases by third parties be regulated? Should academics have different rights from companies?
- Are the provisions of the Data Protection Act adequate in the cae of genetic information and biological sample collections?

Ownership There are two common approaches to the ownership of tissue or parts of the body by the individual. The first is that no one has the right to control the usage of tissues or body parts once removed from the individual. The second approach is that individuals should be able to determine what happens to excised parts of their body.

In terms of ownership of body parts by others, the philosophical underpinning of the law is that the exercise of skill and labour will be the basis for acquiring property rights in a 'thing'. Things can be tangibles, such as tissue samples, or intangibles, such as information derived from isolated DNA. The recent case of *R v. Kelly* demonstrates the 'exercise of labour' approach and provides a justification for intellectual property rights in a patient.

Public policy with respect to the patenting of human genes appears to be in a state of confusion. There are important questions which remain unanswered about the social acceptability of the private ownership of gene patents, and the impact this might have on scientific research, innovation and the costs of new medical technologies.

- Is it socially and ethically desirable for parts of the body to be commodified?
- Does a property approach adequately encompass the ethical issues that surround human tissue?
- Is the 'exercise of labour' approach an appropriate basis for gaining property rights in tissue and information that is derived from DNA?

- Is it ethically tenable that individuals may control the taking of tissue from their body, but can have no control over its subsequent use once it has been altered in some way?
- Is it ethically acceptable that individual tissue may be commercially exploited, but the individual does not have any basis for claiming a part of the profits?
- To what extent does the existing IPR regime surrounding the patenting of human genes adequately meet public policy objectives?
- What are the socio-economic consequences of the monopoly ownership of IPR covering human genes?

The commercial exploitation of biological sample collections The field of human genetics research is marked by a very tight linkage between academia and industry. Many publicly funded research projects end up being commercially exploited, and this is encouraged by public policy. However, where tissue samples are donated freely by participants in research studies, there may be objections to the subsequent development of these resources for profit. If the 'gift relationship' were undermined, this might greatly reduce participation in public research.

Even if the potential economic benefits can be used to justify the commercial exploitation of public sample collections, there are still important issues concerning the basis on which this is done.

To what extent does the close involvement of industry in the exploitation of biological sample collections compromise public support for genetic research?

Should the private sector have the right to build large biological sample collections for genetic research or should such research always be carried out in public-private collaborations?

Should commercial access to public sample collections and genetic information always be given on a non-exclusive basis?

The use of, and access to, personal medical information in genetic research

Consent and the use of personal health information At present, medical records are often used in research without the consent of the individual. Ethically and more recently legally, this has become problematical. Furthermore, one of the important issues in the use of medical records is that they have been created to facilitate the treatment of the individual financed by the public purse. It is increasingly likely that this information will be used in collaborative research with privately funded companies, who will profit out of this information.

- Should (anonymous) medical information be routinely used in research without obtaining consent?
- Who can have access to medical records?
- On what basis should access be granted to third parties?
- Can medical information be used for commercial gain or should it only be used for research?

Confidentiality, privacy and data protection Many of the issues concerning the confidentiality of medical records are similar to those concerning tissue sample collections. The creation of large electronic databases poses problems about security of information and how data should be protected against unauthorized use. Aspects of this are covered by the Data Protection Act. An important issue in this area concerns the way in which doctors can often be used as a 'Chinese wall' to protect the confidentiality of the research subject. However, it depends on the integrity of the doctors concerned, the nature of the professional guidelines, and the clear separation of medical records from genetic information (genotypes).

Are the professional codes of conduct that govern the handling of medical records during research adequate to protect the medical professionals, as well as the patients involved? Who is responsible for ensuring the confidentiality of personal information? Should there be additional oversight mechanisms?

The objectives and findings of research

The social acceptability of genetic research As illustrated by the case studies above, genetic research using large biological sample collections is potentially highly controversial. It is therefore important that the social benefits of research outweigh the risks to society and that the research objectives are socially and ethically acceptable. Without this, there is a danger that the legitimacy of all work on human genetics will be called into question. As discussed above, in research projects based on the study of particular groups, there may also be a need to develop new policies to ensure community consent, in addition to individual consent.

- How can public debate on the aims of genetic research be organized to ensure that participants are well informed and fully involved?
- What mechanisms, such as consultation exercises and citizens juries, are best suited to build social consensus and ensure that broad public concerns are reflected in official policy?
- How can the concept of community or group consent be put into practice in the UK?

The regulation, oversight and governance of research In addition to ensuring that the aims of research are acceptable, it is also essential that the conduct of research commands broad public confidence. This requires a properly functioning system of regulation and research oversight that operates according to clear ethical and legal principles, is transparent and involves a wide range of social groups.

The lack of primary legislation and comprehensive case law in the UK regulating medical research is surprising to the lay person. With the expansion of potentially sensitive and harmful genetic research, the case for more explicit legal regulation is strengthened.

While the established system of research ethics committees has worked well for conventional medical research, there may be a case to strengthen research oversight in the case of genetic studies. Furthermore, there needs to be clear social and legal sanctions to police the system of oversight and to ensure that research is undertaken at the highest ethical standards.

The regulation of sample collections and genetic databases held in the private sector is regulated in largely the same way as public research. However, in the light of the debate about deCODE Genetics, there may be concerns that the commercial secrecy that surrounds the use of personal genetic information within private companies would make effective monitoring and oversight difficult.

- Is the existing legal framework adequate to regulate genetic research in the UK?
- Should the conduct of genetic research be formally regulated by legislation?
- How can research oversight be strengthened?
- Should there be a national bioethics committee that helps formulate policy and provides training for LRECS and MRECs?
- How might families and communities be involved in the oversight of research?
- What additional policies might research funders adopt to ensure that research is conducted to the highest ethical standards?
- Should there be additional safeguards to oversee the development and use of private/company sample collections and genetic databases?

Commercial benefits of research findings Many of the resources being used in genetic research have been developed using public funds and require high levels of public involvement and support (donation of samples etc.). While policy aims to encourage the commercial exploitation of publicly funded research, there is also a recognition that this must be matched by a suitable social return. In the case of research using donated samples, it is generally assumed that the individual donor is giving their tissue to further the collective good of the community, rather than the private profit of a company. However, as has been made clear in several of the case studies, in the future biotechnology and pharmaceutical companies are likely to commercially benefit from collections of tissues donated for research.

This raises important questions about who should pay for and who should gain from research involving biological samples. In addition, it creates the potential for conflicts of interest within the research community.

- On what basis should large sample collections, which may yield significant commercial benefits to industry be established? Should public-private consortia be created?
- How should these collections be financed and how should the financial benefits arising from them be shared?
- How should the commercial exploitation of publicly funded sample collections be governed to avoid conflicts of interest and to protect academic freedom?

Feedback of information to individual participants Genetic research raises a number of issues about the disclosure of information derived from an individual's tissue sample. It would be a breach of confidentiality if a doctor or researcher did disclose the results of a test to other family members. Although, this may be legally correct, it raises difficult ethical questions, as the investigator could have knowledge about a health of a family member which that individual did not possess.

Members of the medical profession involved in caring for patients/participants may also have a legitimate interest in knowing the result of any genotyping. Similarly, other third parties, such as insurers and employers may find genetic information of commercial benefit to them.

- Do research subjects have a right to know information that affects them?
- Do purely anonymous studies run counter to safeguarding individual health?
- Do participants also have a right not to know?
- Should information be fed back to genetically related non-participants?
- Should it be passed onto the participant's doctor in situations where their health care could be improved?

- What access to information should be given to other interested third parties?

Potential socio-economic impacts of research One of the most important questions for policy makers and social scientists is trying to assess the potential impacts of this type of genetic research. These might be felt in a number of areas, including, health care, personal identity, health policy, discrimination and in the competitiveness of UK industry. There will be a large amount of diagnostic information about the association between having a particular genotype and getting a specific disease. However, diagnosis will be available long before effective therapies are introduced in many cases. If this information were widely available, it could lead to the creation of new classes of 'patient' (the asymptomatically ill or disabled) for whom little could be done. The widespread use of genetic information would also have important implications for health policy and the emphasis given to drug-based therapies, perhaps at the expense of environmental protection and social improvements.

Much discussion has already taken place about the potential for discrimination which the introduction of new genetic screening and diagnostic technologies might present in the areas of employment and insurance. Functional genomic research might significantly increase the possibility of this occurring. However, it must be stressed that significant clinical and commercial benefits are likely to result from research involving sample collections. The public policy objective must therefore be to regulate the creation and dissemination of genetic information in such a way that the maximum benefits for health care and industry are ensured, while protecting research subjects and civil rights.

- What is the utility of predictive genetic information in cases where no therapy exists?
- What are the implications for health and public policy of changing concepts of the body and illness causation?
- How real is the potential for discrimination based on this new genetic knowledge? What measures can be adopted to protect people's civil rights?

- How can the application of new genetic knowledge in health care and its incorporation into new technologies be regulated in such a way as to strengthen the development of the UK biotechnology industry, while minimizing the adverse consequences for society?
- Finally, what is the future for collective health care provision in the UK in the context of genetics research which is intrinsically concerned with individual factors?

Acknowledgements

This is an edited version of a background paper written for a national workshop on 'The Collection of Human Biological Samples for DNA and Other Analysis' organized by the Wellcome Trust in London on the 5th November 1999.

The authors would like to thank the following people for their time and support in completing the original paper for the Wellcome Trust: Dr Tom Wilkie, Dr Pat Spallone, Dr Frances Rawle, Dr Pat Goodwin, Dr Diana Chase, Dr Trevor Nicholls. Special thanks must also go to Jayne Dennis, a summer intern at Nottingham, who carried out a comprehensive literature search for the chapter, and Pat Hulme, who provided much needed administrative support.

Notes

1. No. 139/1998 passed by Parliament 22 December 1998, at the 123rd session 1998-99.
2. www.decode.is
3. *New Scientist*, 5 December 1998.
4. *New York Times*, 23 January 1999.
5. Interview with David Thor Bjorgvinsson, Chairman of the Monitoring Committee on 6 July 1999.
6. Anon. (1998) MRC funds large-scale human genetic database, *Nature Medicine*, 4 (12), p.1346.
7. See Anon (1999) People power: population profiles and common diseases, *Wellcome News*, Q3, p.18.
8. *Ibid.*
9. Animals (Scientific Procedures) Act 1986.

10. Convention for the Protection of Human rights and the Dignity of the Human Being with regard to the Application of Biology and Medicine (Convention on Human Rights and Biomedicine) 1997 ETS No.164.
11. Visit to the Council of Europe web site: http://www.coe.fr/tablconv/164.htm (21 September 1999).
12. Medical Research Council (1999) *Report of the Medical Research Council Working Group to develop Operational and Ethical Guidelines for Collections of Human Tissue and Biological Samples for Use in Research*, Third Working Draft.
13. Royal College of Physicians Committee on Ethical Issues in Medicine (1999) Research based on archived information and samples, *Journal of the Royal College of Physicians of London*, 33 (3).
14. Royal College of Pathologists (1997) *Consensus Statement of Recommended Policies for Uses of Human Tissue in Research, Education and Quality Control* (London: Royal College of Pathologists).
15. Office of Technology Assessment (1987) *New Developments in Biotechnology: ownership of human cells and tissues* (Washington, DC).
16. *Chatterton and Gerson* (1981) 1 All ER 257.
17. Dyer, C. (2000) GP reprimanded for testing patients for HIV without consent, *British Medical Journal*, 320, p.135.
18. Kennedy, I. and Grubb, A. (1998) *Principles of Medical Law*, Oxford: Oxford University Press.
19. s.1(2) Human Tissue Act 1961.
20. Dyer, C. (1999) Government orders inquiry into the removal of children's organs, *British Medical Journal*, 319, p.1518.
21. s1(1) Anatomy Act 1984.
22. *Herbert* (1961) 25 J Criminal Law 163.
23. *R v. Welsh* (1974) RTR 478.
24. *R v. Rothery* (1976) RTR 478.
25. (1998) 3 All ER 741.
26. Lord Lester of Herne Hill QC appearing for the General Medical Council in the *R v. Dept. of Health ex parte Source Informatics, The Times,* 18 January 2000 (Transcript Smith Bernal 21 December 1999).
27. *W v. Edgell* (1990) 1 All ER 835.
28. *The Times*, 18 January 2000 (Transcript Smith Bernal 21 December 1999).
29. Department of Health (1997) *Report on the Review of Patient Identifiable Information*.
30. s.33(1)(b) Data Protection Act 1998.

PART III
GEN-ETHICS: HUMAN GENETIC BANKING

7 DNA Sampling and Banking in Clinical Genetics and Genetic Research[1]

Kåre Berg

Introduction

Long-term storage of biological material containing DNA and data pertaining to stored biological material is by no means new. At least in Norway, pathology departments have stored tissue samples from autopsies and biopsies to the extent that tissue samples from about 6 million persons reside in the pathology departments of a nation with only 4.5 million inhabitants. Data on the persons that the tissue specimens originate from are also stored in hospital records and in departments of pathology. Samples or data are not anonymized, and there is no feeling within the profession that names should be removed. Not infrequently, the pathologists re-examine specimens: for example, if there is new diagnostic information that may cast doubt over the original diagnosis. Important new findings could be made, for example, if new microscopic slides are prepared from a stored tissue block and stained with other stains than originally used or exposed to various, labelled antibodies.

There is little doubt that tissue specimens stored and labelled with identifiers are potential tools for attempts to improve quality control. Any initiative to have collections of tissue specimens or corresponding data destroyed for data protection or other purposes is in conflict with the need for quality control of diagnostic laboratory work. Furthermore, destruction of collections of tissue specimens would amount to destruction of potentially important research material. Destruction should only be contemplated if severe damage could result from continuing to store biological specimens; trivial reasons should not lead to destruction of samples that could be used for research.

Importantly, there has hardly been any move previously to destroy stored biological material in departments of pathology. Is it the new, perceived 'mystery' of DNA that is changing public attitude? Genes were present in tissue specimens also in the days when nobody paid attention to genetics, and functional genes in cells have for a long time been passed on from individual to individual through blood transfusion or organ transplantation.

Biobanks

Terms such as 'biobanks' are used for collections of biological specimens. Even collections of data from laboratory analyses are sometimes considered as 'biobanks'. I prefer to use the term 'biobank' about collections of actual biological material that one stores ('banks') in order to be able to make use of the specimens at a later stage. Terms such as 'genetic banking' may be appropriate for collections of biological material that have been assembled for purposes of genetic investigations (for example, in connection with health problems in a family) or are particularly suitable for genetic studies. The most interesting 'genetic banks' would be those with living cells that can be further cultured or where PCR studies and other genetic analyses can easily be done. Collections of samples from extended families with various disorders are particularly useful for genetic linkage studies to identify genes responsible for human diseases.

Some characteristics of genetic disorders

Genetic disorders have certain characteristics that are highly relevant to the banking of actual biological material as well as of data from genetic analyses:

1. The disease can be diagnosed with a very high level of specificity by the study of DNA from affected persons and

their family members. Biological material stored under optimal conditions or even as tissue blocks fixed for microscopical analysis can be used for diagnostic work, long after the death of the patient.

2. Knowledge of disease status in one individual will entail information about disease risk and prognosis of other persons (relatives).
3. Health information about persons now alive has implications for the health of people in future generations.

These characteristics call for the storing of very precise and detailed information concerning test results with a view to helping new members of the kindred who want their genetic status clarified. The information should be permanently stored, since it will be of importance for generations not yet born.

Uses of 'genetic banks' in medical genetics services

As research in clinical genetics moves forward and the Human Genome Initiative approaches its completion, the prospect for uncovering genetic causes for disease improves significantly. Stored biological material will make it possible to apply new knowledge to the study of DNA of people who are no longer with us, from families with aggregation of persons with disease in a manner that is suggestive of a genetic basis. Having identified the molecular cause of disease in persons no longer alive, one would be in possession of a tool to examine the health status of living persons, evaluate their disease risk, and institute preventative measures as appropriate for the individual's genetic status.

Stored biological material as well as stored results of laboratory analyses are of great practical importance, also for work with disorders that are well known today. For example, a young adult with borderline cholestorol level whose father died from a premature heart attack at age 50 or 55 could represent a diagnostic problem. In young adults, the mere measurement of cholestorol may not be adequate to conclude that a person has a mutation in the low density lipoprotein (LDL) receptor (LDLR) gene [LDLR mutations cause classical familial hypercholesterolemia with risk of early coronary heart disease (CHD)]. If an LDLR mutation were present in the father but not in the daughter or son, the prognosis would be better than if the mutation had been passed on and attempts to reduce moderately increased cholesterol in the young person by dietary changes only would be justified.

Stored data alone would have been helpful if the person's father or mother had a known LDLR mutation. However, most test panels for LDLR mutations detect only a proportion of the errors in this gene and, in the situation alluded to, one would want to conduct an extensive search for LDLR mutations if no such mutation had been found in the parent who had died from a premature heart attack. To do an extended search would require stored biological material. Perhaps the most important outcome of testing our young person would be to find out that he did not have the same LDLR mutation as the deceased parent and would therefore not have to have life-long drug treatment. Thus, genetic banking may be crucial to choice of treatment.

In general terms, 'genetic banking' represents the possibility of comparing the genetic status of a living patient with that of a person who has died, as well as to compare DNA patterns between living people.

'Genetic banking' and research in medical genetics

Banking of biological material from deceased persons makes it possible to include exact genetic information on such individuals in genetic research, be it in the area of genetic linkage studies, molecular epidemiology, study of the natural course of disease or pharmacogenomics. Stored biological samples from families with aggregation of cases of disease make it possible to search for co-segregation of disease with numerous genetic markers to uncover the site in the genome of the gene(s) causing or contributing to the disease. In due course, it would then be possible to decide which gene in the region would be the likely cause of the disease. Knowledge of the exact nature of genes that cause diseases and their normal counterparts could pave the way for new treatment modalities. Population analysis of genes that cause disease will uncover gene frequencies and other characteristics of genes in populations.

Comparison between groups of patients with different mutations in a clinically relevant gene could lead to identification of distinct clinical pictures, specific to different mutations in the same gene. Different mutations (in the same or different genes) may well turn out to require different therapeutic modalities.

Genetic variation that determines people's reactions to various drug treatment are already known. Increasing knowledge in this area of pharmacogenomics will probably lead to fewer side effects of drug treatment because choice of drugs and dose of drug will be tailored to the individual's need. The efficiency of drug treatment will improve. Persons who need higher than average doses, for example because of increased drug metabolism, may be systematically identified prior to drug treatment. In a few years' time, it is likely that many drugs will come with instructions for use that differentiate between genotypic groups, and genetic testing may precede drug treatment. Furthermore, recorded reactions to drug treatment in people who are no longer alive could be correlated with genetic variants in different enzyme systems if biological materials were stored and available for testing.

Genomic research recently led to the discovery by four different groups of workers of the basic defect in so-called Tangier disease. This turned out to be a mutation in the so-called ABC-1 gene. The mutation reduces lipid efflux from cells. One of the characteristics of Tangier disease is a very low level of high density lipoprotein (HDL) cholestorol (HDLC) (often referred to as the 'good cholestorol') in serum. Most importantly, other mutations in the same gene were found in kindreds with coronary heart disease (CHD) where the patients had low HDLC level. Thus, information arrived at from the study of a very rare disease, turned out to throw light over the genetic basis and disease mechanisms for a very common disorder: CHD. A vast number of biological specimens from patients who had low levels of HDL and who died from premature cardiovascular disease are stored in hospitals. Samples from such persons could be examined with respect to mutations in the ABC-1 gene, and any mutation observed could then be searched for in living relatives.

Those having a mutation would be in need of preventative efforts. Furthermore, it seems plausible that this new knowledge about a mechanism contributing to disease in many patients with low HDLC level, may lead to efficient correction of the faulty lipid efflux from cells, and a breakthrough could come soon. If this were to happen, a vast number of people could be helped by this research.

Thus, 'genetic banking' will facilitate important research progress.

Ethical problems in 'genetic banking'

Although the storage for several decennia of biological specimens has in general not led to major problems, there are ethical problems relating to the storage of specimens as well as test results. However, the above and other advantages of 'genetic banking' outweigh the disadvantages, in my opinion. The need for storage of samples and test results is long term for medical genetic services (genetic diseases transcend generations) and for genetic research. This long-term perspective calls for strict rules and practices with respect to the protection of data and specimens. Several problems related to 'genetic banking' should be considered and ethically acceptable solutions should be found. Some of the problems are discussed below.

Sampling of DNA for 'genetic banking'

The drawing of blood for extraction of DNA for diagnostic purposes and genetic counselling should not lead to significant ethical problems if test results and specimens are fully protected from third parties. However, the persons whose blood is drawn should be made aware of the fact that samples and data will be stored and possibly used for additional tests, as new knowledge may make this desirable, and that more testing may be conducted for purposes of quality control. If applicable, people should also be informed that the genetic centre may want to use a person's DNA also for research purposes. The person's permission has to be asked for long-term storage of DNA as well as for research, following an information process that encompasses all foreseeable consequences of testing, sample storage and research. The person's consent should be as informed as her or his education and level of knowledge permit. For purposes of research, it is important to stress the voluntary nature of participation and that refusal to participate will have no negative consequences with respect to the person's relation to the healthcare system.

Narrow or broad consent

In some countries, including Norway, there is a tendency for data protection officials and research ethics committees to demand that the informed consent given should be very narrow and that specimens obtained should only be analysed with respect to variables explicitly mentioned at the outset of a research project. If it becomes desirable to study other variables, such as single nucleotide polymorphisms (SNPs) in other genes, the demand is that the researcher should re-contact the person from whom the sample originates and ask for an additional, narrow consent, and this could presumably happen again and again. Some research ethics committees would even decide that nothing more could be done to a sample if the person is no longer available to give new consent. However, other research ethics committees may willingly give permission to expand a study.

A demand for narrow informed consent can be a serious obstacle to important research. It may make it impossible to search for genetic markers of mortality, or to go back to an existing collection of specimens to test a new hypothesis, perhaps originating from knowledge that was not available when the samples were collected. Repeated consent would also make research more expensive, to the extent that it may not be conducted. Finally, it could foster negative attitudes in the population if researchers repeatedly had to come back asking for new consent.

In real life, people are rarely afraid to give broad consent, and researchers frequently hear that participants in studies want as much research as possible to be done on their DNA or other samples. In my view, it infringes the autonomy of a person if ethics committees, data protection officials or law-makers should deny a person the right to give a broad permission for research, following adequate information. A rule demanding very narrow informed consent is paternalistic and does not reflect a strong wish in study participants, *provided* test results and specimens are fully protected from third parties. I think the burden of evidence lies with those who want to maintain a demand for narrow consent, and they ought to be able to argue convincingly why people who want to provide broad consent should not have the right to do so.

The discrimination issue

The most serious ethical issue in connection with 'genetic banking' is the risk of discrimination against tested persons (or for that matter persons who refuse genetic testing). This very serious area has been debated in as different fora as the political leadership in the United States, the Council of Europe and the Norwegian Parliament. There is broad agreement that it cannot be tolerated that a person is discriminated against because of her or his genes. To prevent discrimination is a matter for law-makers. In Norway, this has led to a law forbidding insurance companies and other institutional third parties to make use of or ask for results of genetic tests, but the situation is complicated because insurance companies have the right to ask health-related questions. These could be asked in a way that the answer might suggest a genetic disease risk. Also, there are a handful of examples of attempts on the side of insurance companies to act in conflict with the law.

To the best of my knowledge, most countries do not have a law forbidding third parties such as insurers or employers to make use of genetic information, although an understanding between the insurance industry and the government has been reached in some countries, including Sweden.

There is no tradition for providing insurance companies, employers or others with medical information on individuals participating in a research project. Therefore, the risk of discrimination on the basis of results of genetic tests mainly occurs in a diagnostic situation, often related to genetic counselling, or in situations where insurers or others outright ask for test results or for tests to be conducted. The risk of discrimination may not be limited to persons consulting the physician/geneticist or applying for insurance or employment. Because of the very nature of genetic diseases, testing of one person may result in information about disease risk in other persons (relatives) who may not have addressed a genetic centre, an insurance company or an employer. Such persons may not even know that unauthorized information about them exists, for example in an insurance company, because of a relative's application for insurance. I see no other way to protect consultants and their relatives against discrimination other than diagnosis, treatment and prevention of disease, and related research. The use of genetic tests in criminal cases, paternity testing or in connection with family reunion would be exceptions.

'Ownership' of samples in 'genetic banks'

At face value, it may seem plausible to state that the 'owners' of biological specimens in 'genetic banks' are the persons from whom the samples originate. It may also seem plausible that these persons retain a right to decide what could be done with the samples and that they should benefit financially if any revenue ascribable to the samples would ever accrue.

However, the situation may be more complicated. The genetic centre that has extracted DNA and conducted analyses has 'invested' in the sample in fact to a greater extent than the person from whom it originates. This must give a genetic centre some rights and it would seem unfair if a person were to withdraw a sample given for research, once the researchers have invested time and efforts in the sample. It is not even clear that the donor of a sample is *entitled* to share in any financial gain that may result from study of his DNA. It is in fact the researcher's analysis of the sample that makes it 'valuable'. The donor of the sample would not in an ordinary way be entitled to 'royalties' or payment for 'licensing'. The donor has not made a contribution to the 'value' that the sample may acquire. His situation is not similar to that of a holder of intellectual property or a patent. The Ethics Committee of the Human Genome Organization (HUGO) considers the genome as part of the common heritage of man (HUGO, 1996). This attitude does not support the notion that the donor of DNA 'owns' the sample.

On the other hand, the donor must have some rights if important research progress is made on the basis of her or his DNA. This would seem particularly obvious if third parties have significant financial gains on the basis of a person's DNA. A major difficulty, however, would be that a great number of years may elapse from when the sample is drawn until financial gain accrues (for example in the pharmaceutical industry following development of new drugs).

The issue of 'ownership' to samples in 'genetic banks' remains unresolved and needs more study and discussion, hopefully leading to consensus. It may not be fruitful to consider samples that have been prepared and extensively examined in a research project as the donor's property in a traditional sense.

Work in the ethics field within the Human Genetics Programme of the World Health Organization

The Human Genetics Programme of the World Health Organization (WHO) has worked on ethical problems related to research in medical genetics and genetic services for many years and issued a major discussion document in 1995 that has been widely read and commented upon. Based upon this background document, comments received from many quarters and discussions all over the world, unanimous decisions were reached on numerous ethical questions in a WHO expert meeting with participants from all regions that took place in December 1997. The final text of WHO's 'Proposed International Guidelines on Ethical Issues in Medical Genetics and Genetic Services' was issued in April 1998 (WHO, 1998).

The major ethical principles that form the fundament for the proposed guidelines were:

- autonomy: respecting the person's autonomous decisions;
- beneficence: giving the highest priority to the welfare of persons, and maximizing benefits to their health;
- non-maleficence: avoiding and preventing harm to persons or at least minimizing harm;

- justice: treating persons with fairness and equity and distributing the benefits and burdens of health care as fairly as possible in society.

Each recommendation in the proposed guidelines was based on one or more of these basic principles.

With respect to banked DNA, the WHO expert group identified several points to consider, including:

- protection of individuals from possible discrimination by employers, insurers and others;
- possible benefits to the individuals from research findings;
- possibility of multiple uses of the same sample in different and even unforeseen research projects;
- possible sharing of samples;
- advantages and disadvantages for individuals and researchers of removing all identifiers from a sample.

The WHO expert committee offered several points of guidance with respect to banked DNA, including the following:

- 'Blanket' informed consent should be preferred since this would make multiple uses of samples in future projects possible.
- Control of DNA should be familial, not only individual.
- Blood relatives may have access to stored DNA for purposes of learning their own genetic status but not to learn the donor's status.
- DNA should be stored for as long as it would be of benefit to present or future relatives.
- Attempts should be made to inform families at regular intervals of new developments in testing and treatment.
- There should be no access for institutional third parties to banked DNA, except for forensic purposes or instances where the information is directly relevant to public safety.
- Insurance companies, employers or other third parties that may be able to coerce consent should not be allowed access to banked DNA, even with the individual's consent.

- Potentially valuable specimens that could be useful to family members in the future should be saved and made available.

Only the specific recommendations concerning banked DNA where there was unanimous agreement are included in the document from the WHO expert meeting.

Conclusion

Banking of biological specimens for the purpose of genetic analysis is useful for diagnosis, prognostication, therapeutic consideration and preventative efforts in medical genetics services, and offers possibilities for research that may lead to important new progress, including new possibilities for treatment of serious diseases. In countries without laws prohibiting use of genetic tests of genetic information by third parties such as insurers or employers, the risk of discrimination could be an obstacle to use of 'genetic banks' for medical genetics services. Local rules for protection of data and samples may have to be developed and applied until adequate laws have been established. The potential disadvantages, once full protection of test results and samples are secured, are small compared with the advantages to persons, families and society in general.

The usefulness of 'genetic banks', particularly for research purposes, would be markedly reduced if authorities demand only very narrow informed consent. If protection of test results and samples are adequately taken care of, there may be no significant need for narrow informed consent. Demands for narrow informed consent do not reflect the prevailing attitudes of most people participating in research projects but rather a paternalistic attitude in opinion-makers. A demand for narrow consent violates the autonomy of study participants. People should have a right to volunteer for as much research as they want.

The need for banking of test results and samples is long term, and DNA samples should not be destroyed as long as they could be useful for genetic services or research. The need for long-term storage calls for strict protection of data and samples, preferably in medical genetics centres. The question of 'ownership' to samples and data in 'genetic banks' is complicated and needs more study and discussion.

Note

1. ESRC Seminar on Human Genetic Banking, Centre for Professional Ethics, University of Central Lancashire, Preston, UK, 10 February 2000.

References

Human Genome Organization (HUGO) Ethics Committee (1996) statement on the principled conduct of genetic research, *Genome Digest*, 3, pp. 2-3.

World Health Organization (1998) Proposed international guidelines on ethical issues in medical genetics and genetic services, *Document WHO/HGN/GL/ETH98.1*, Geneva.

8 Human Genetic Banking and the Limits of Informed Consent

Garrath Williams and Doris Schroeder

Introduction

The frequency and scope of human genetic banking has increased significantly in recent years and is set to expand still further. Here we consider how the key ethical considerations raised by these developments are most usefully framed. Our focus will be upon medical and population research. That is to say, we shall leave to one side a related development of potentially still greater import, that of forensic databases, which promises greatly increased state power to investigate individuals' activity, criminal or otherwise.

There is, however, an interesting disanalogy between these two sorts of databases, in terms of the way ethical discussion tends to be structured. In the case of forensic databases, discussion is overwhelmingly structured by an opposition between the public good and individual rights: the public interest (in detecting and even preventing crime) and civil liberties or privacy considerations, that militate against large-scale state access to genetic samples. In the medical context human genetic banking has attracted a quite different discourse, with individual rights at the forefront: those relating to the informed consent of actual and potential donors to the database. So far as the public good goes, this has tended to be addressed only by rather weakly supported claims regarding possible health care benefits.

It is not surprising, then, that some recent contributions hint that informed consent is being asked to play too great a role in our thinking about the ethics of these projects. In this chapter, we make such a case explicitly. We stress how limited the informational aspect of 'consent' must be for most donors; and suggest that much more searching scrutiny is required of the organisation and exploitation of these projects, particularly in the light of the commercial forces at work here. We argue that the individualistic focus of informed consent must be supplemented by pro-active forms of on-going institutional governance, and that this requires greater attention to the duties pertaining to the custodians of gene banks, and to the institutions that fund them.

We structure our discussion into three main sections: a brief consideration of why scientific research now demands large human gene banks; the question of informed consent, and issues which that raises (for instance, confidentiality and impact on third parties); and how to conceptualise the moral issues at stake, given the limitations of a framework oriented solely by informed consent.

Human genetic banking for pharmacogenomics and population genetics

By human genetic banking we mean large-scale banks which contain either tissue samples, from which genetic material might be or has been extracted or genetic information, which may be coded and stored in various forms; and, *in addition*, health and 'lifestyle' information pertaining to the sample donors.

Two current research areas require human genetic banking in previously unknown dimensions: pharmacogenomics and population genetics (Bell, 2000, p.2; House of Lords, 2001, chapter 2, p.1).

Pharmacogenomics relies on the hypothesis that responses to medication are significantly affected by a patient's genetic make-up. Hence, the aim of research is to ensure 'the right medicine for the right patient' (Wolf, 2000, p.1). The advantages of tailor-made medication are obvious. Patients are not put through futile therapies, or therapies likely to cause them dangerous side-effects; medication is not wasted on those who cannot benefit from it. To correlate genetic make-up with reactions to drugs is, however, a matter of 'statistical brute force' (Lowrance, 2001, p.1009). According to the Medical Research Council (2000, p.5), most collections existing world-wide 'are too small to allow statistically meaningful research'.

Population genetics relies on the hypothesis that common diseases can be linked to a complex interplay of genetic predispositions with life style and environmental factors. Researchers are seeking to identify susceptibility genes for diseases such as early-onset heart disease, asthma, depression, osteoarthritis, migraine, Alzheimer's, Parkinson's and diabetes (Glaxo Wellcome, 2000, p.2). Ideally, it will become possible to give tailor-made advice on life-style and environmental issues to people at risk from particular diseases.

In addition, new drugs might be developed if the links between genetic make-up and such diseases can be understood.

Human genetic banking – three examples[1]

The best known human genetic database suitable for pharmaco-genomics and population genetics research is the Icelandic collection of the country's health records. The Icelandic Parliament granted a licence to deCode genetics in December 1998. The database contains tissue samples and identifiable health data from official medical records and is based on 'opt-out' consent (records are included in the database unless a citizen takes specific measures requesting non-inclusion).[2] Drawing on a population of 270,000, the database offers enormous potential for genetic research: detailed medical records of every Icelandic inhabitant have been kept for over fifty years, as have tissue samples for many; genealogical data also exists for most Icelanders (cf. Chadwick, 1999, p.442).

An even larger database is about to be set up in Britain where the Medical Research Council in co-operation with the Wellcome Trust and the National Health Service (NHS) are planning 'BioBank UK' (previously referred to as the 'UK Population Biomedical Collection'). DNA and lifestyle and medical information will be sought from approximately 500,000 people between 45 and 64 years old (the average time of onset of several common diseases). Samples will be held in public ownership, and data will be made available to commercial companies (Wellcome News, 2000, p.11).

A much smaller British database is already in place, the 'North Cumbria Community Genetics Project'. The database stores umbilical cord blood and tissue from newborn babies and their mothers, in addition to limited personal data (about 5,000 samples by 2000). As there is little population movement in the area, long-term follow-up of clinical outcomes is envisaged. The project was triggered by local concerns about genetic implications of the proximity of British Nuclear Fuels at Sellafield (North Cumbria Community Genetics Project, 2000). Samples and data are taken only following explicit consent by the mother, and the whole project was preceded by extensive community consultation (North Cumbria Community Genetics Project, 2000, p.3 and 2001a).

What the above three banks have in common is that they provide or will provide extensive, centralised, and perhaps even standardise genetic information, together with clinical and personal data. It is the ongoing link between genetic and clinical information that makes these banks suitable for pharmacogenomics and population genetics research, as well as their size. In both fields, the linking of genetic (biological) and clinical (personal) data is essential (Martin, 2000, p.3). Total anonymisation[3] is of course possible, but would prevent the introduction of health data obtained *after* anonymisation; in other words, the *prospective* character of the database would be lost. Arguably, this would compromise its scientific value considerably (Glaxo Wellcome, 2000, p.11; Medical Research Council, 2000, p.10).

Informed consent and human genetic banks

Currently the most prominent ethical issue concerning genetic banking relates to individuals' consent for the use both of their tissues and their health care information. Informed consent may be defined as:

> The knowledgeable and voluntary agreement by a patient to undergo an intervention by a health-care professional, one that is in accord with the patient's values and preferences.
>
> (Moreno *et al.* 1998, p.687)

The information side of 'informed consent' or 'knowledgeable agreement' requires the explanation of a medical intervention's purpose, its potential benefits and foreseeable risks, as well as its alternatives – all this in a way that is intelligible to a patient or volunteer (Brody, 2001, 11f). The 'consent' aspect requires a noncoercive setting to obtain agreement as well as some form of authorisation or documentation. (Clearly, other principles will be relevant as regards emergencies, and the situation of incompetent adults and young children.)

How pressing the issue of informed consent has become in the storage of human biological materials has been amply shown by events at the Alder Hey hospital. According to The Royal Liverpool Children's Inquiry (2001), the hospital did not comply with the Human Tissue Act (1961, Section 1(2)) when removing, retaining and disposing of organs and human tissue from deceased children. Instead, it built up one of the most extensive collections in the world of children's hearts as well as substantial collections of other body parts. In doing so without consent, considerable lack of respect was shown toward parents, who were to discover years later that their child's body had been stripped of some or even all internal organs, and been buried as a 'shell' (Department of Health, 2001).

What aspects of human genetic banking are problematic for the informed consent procedure? Clearly, some issues are more than usually straightforward: the absence of immediate therapeutic benefit is clear, as is the negligible risk posed by current medical procedures for taking samples. The fact that prospective donors are usually healthy volunteers means that consent for research is given under less threatening and distressing circumstances, as compared with hospital settings (where, for instance, a patient suffering from a debilitating illness would be more vulnerable to excessive deference to doctors or feelings of intimidation). Nonetheless, there are two especial sources of difficulty: the linking of genetic and clinical data, which in turn poses issues around confidentiality, withdrawing, feedback and recontacting; and that of third party involvement.

Linking genetic data and clinical data raises significant problems for informed consent. As we have seen, in both pharmacogenomics and population genetics, the integration of genetic and clinical data is deemed essential, and donors' health care data is to be input into the bank on an on-going basis. This generates issues about withdrawing, feedback and confidentiality, which do not arise in such a general form in other medical research.

The possibility of *withdrawing* from a research study is one of the basic principles regulating medical experiments according to both the Nuremberg Code (1947, Standard 9) and the Declaration of Helsinki (1996, Basic Principle 9). It is essential that informed consent procedures explain this basic right to subjects of research in pharmacogenomics and population genetics (cf. Porter, 2000, p.53, p81). The North Cumbria bank makes provision for complete withdrawal, so that all data and samples are removed on request, while the Icelandic one has a complicated and arguably unsatisfactory opt-out option (cf. Rose, 2001, p.25); regulations for BioBank UK have not yet been finalised. Neither the Icelandic Health Sector Database nor the North Cumbria Community Genetics Project make it possible to 'opt into' certain research projects and 'opt out' of others. Either a person withdraws altogether, or the records can be used for all projects approved by the relevant ethics committee (North Cumbria Community Genetics Project, 2001; Haraldsdottir, 2000, p.12).

The genetic component of research into pharmacogenomics and population genetics can offer no therapeutic benefit to volunteers. However, research aims include the identification of people who might or might not benefit from certain medications and who are at risk of developing certain common diseases. Volunteers who understand these objectives might reasonably expect to be given such *feedback* on the analysis of their samples. This is rightly considered a particularly 'tricky issue' amongst human genetic bankers (Wellcome News, 2000, p.11). With the North Cumbria Community Genetics Project (2001a) and the Icelandic Health Sector Database, in general no feedback is given to individual volunteers. The draft protocol for BioBank UK likewise says that no feedback will be provided (Wellcome Trust, 2002: p.5.4.1). There is certainly a major resource consideration here: some have even argued that regular feedback would transform a research project into genetic screening requiring an enormous budget for genetic counselling of all donors (e.g. King, 2000, p.4). Clearly, given that future developments are so uncertain, there are grave problems in foreseeing what implications feedback might have, both for particular individuals and the management of collections.

The Medical Research Council (2000, p.10), however, expects feedback to be given on a regular basis and therefore recommends that banks budget for genetic counselling. It might also be noted that the UK study on 'Public Perceptions of the Collection of Human Biological Examples' found that most people thought donors should have 'the right to feedback on anything that emerged from their own sample' (Porter, 2000, p.8). This might plausibly be seen as something owed "in return" for a donor's participation. Regardless of which position one supports, so far as informed consent is concerned the crucial point is that research subjects know in advance whether they will receive feedback on their samples (Chadwick, 2001, p.207; Clarke *et al.* 2001, p.90ff), though just what sort of information may turn out to be available cannot yet be known.

Another major issue concerns *what type of consent to obtain*, which is bound up with the question of whether research subjects should be recontacted in the case of new studies. Evidently, the collections we have mentioned are suitable for a wide range of research studies. Should donors, when depositing a sample, be asked for narrow or open-ended consent? This could range from consent for material to be used in one, well-defined research study (so that future uses would require new informed consent procedures) to open-ended consent for general research (with the main safeguard against 'undesirable' research probably being offered by some relevant research ethics committee). To obtain informed consent from every participant for every research study is, obviously, very time-consuming and inconvenient for researchers, particularly with these large collections. This would imply that some sort of open consent is the only workable option. Interestingly, the recent report on 'Public Perceptions of the Collection of Human Biological Examples' found that most respondents were comfortable with the idea of samples being used for disease-specific work, but not for general research (Porter, 2000, p.7).

One in-between option, consent for general research approved by an ethics committee, with certain further limitations, has been adopted by The North Cumbria Community Genetics Project. Donors, or rather donors' parents, must to a significant degree trust the judgement of an ethics committee but are assured in advance that samples will only 'with extreme caution and subject to rigorous examination' be used for 'studies that involve intelligence, behaviour, personality and psychiatric disorders' (North Cumbria Community Genetics Project, 2000, p.5). Beyond the point of donation, though, donors are unable to influence the direction of ongoing research – except by withdrawing should they disapprove of decisions made by project leaders.

A further complicated issue concerns *confidentiality*, one of the most fundamental and long-standing principles of medical ethics. As we have seen, confidentiality is threatened by the need to make linking access to health care data, if researchers are to incorporate data on an on-going basis. Coding of samples and data can reduce but not eliminate the possibility of breaches of confidentiality. (By contrast a system of complete data anonymisation makes it almost impossible to link any sample to an identifiable individual). This formed a particular concern of GPs and nurses about BioBank UK, 'give the size of the sample and their own experience of maintaining secure records' (Porter, 2000, p.10; cf. AstraZeneca, 2000, p.2).[4]

These worries are surely well-founded: confidentiality breaches can occur both accidentally, and on public health or criminological grounds. In a recent UK case, Scottish police seized materials from a clinical study on HIV (Yirrell *et al.* 1997), in disregard of guarantees of confidentiality given to the subjects when they consented to take part. This led to one subject's conviction for culpable and reckless conduct (knowingly exposing his partner to HIV), and a sentence of five years' imprisonment (Dyer, 2001; Brown, 2001). The Medical Research Council, the funder of the research, claimed: 'Confidentiality has never been an absolute... The case involved an investigation into a serious crime... There was thus a strong public interest in disclosure...' (Dyer, 2001a).[5] Thus breaches of confidentiality and respect are both possible and, on occasion, arguably legitimate. As well as the state, two other powerful parties might want to acquire genetic information about individuals – employers and insurance companies (Clarke, 2001, p.143). Both will want to select individuals who are likely to stay healthy (ibid.), whilst agencies of the state may seek access to genetic information for forensic or 'security' reasons, or in case of paternity suits.

So far as informed consent is concerned, two conclusions suggest themselves: that consent procedures should draw research subjects' attention to the potential for data breaches and their consequences; and that the administrators of databases should clearly indicate when they would be prepared, or could be obliged, to release data.

Interestingly, the North Cumbria project promises donors that samples will 'never be available for... police records' and will 'only be used for medical research' (North Cumbria Community Genetics Project, 2001a). Yet it seems clear that, in the UK at least, no researcher is in a position to offer such a guarantee.

Finally, complex issues arise with regard to possible *effects on third parties*, which need to be explained during the informed consent procedure. The first is commercial third party benefit, which raises issues of distributive justice and possible exploitation. The second is indirect third party harm such as unwanted health knowledge or ethnic discrimination.

Human genetic banks, as required for pharmacogenomics and population genetics, 'will be driven by demand from the commercial sector' (Womack and Gray, 2000, p.251). This pressure may be of mutual benefit to biotechnology companies and communal health care systems, if it results in, for instance, better and cheaper pharmaceutical products (ibid. p.252) or more efficient ways to operate a health service (Department of Health, 2000, p.1; Haraldsdottir, 2000, p.12). If forthcoming, benefits will accrue to the population at large, not, in the first instance, to sample donors. Conversely, in this altruistic or 'gift' scenario, donors will most directly benefit commercial (profit-making) organisations. Research participants need to be made aware that there might well be possibilities for commercial exploitation in addition to any humanitarian benefits (Chadwick, 2001, p.209) and, moreover, should individual samples lead to major new pharmaceutical discoveries, that a share of the profits will not ensue.

Third party harm is another issue, particularly with regard to research in population genetics. 'Third parties' in this context are those who share a similar genetic make-up with the research participants. For example, Glaxo Wellcome's (2000, p.2) programme in population genetics includes research to identify susceptibility genes for Parkinson's disease or Alzheimer disease. It is possible that one member of a family will agree to take part in this research whilst another declines, or for some other reason is not party to the study. In this case, the unasked or unwanted disclosure of genetic information to non-consenting parties could occur, overriding their personal autonomy. No definitive solution to this problem is possible; one might, though, attempt to enlighten the family member who is donating (and, as a result, being tested) of his or her consequent responsibilities toward other family members (ibid, p.68), particularly drawing his or her attention to the complexities of unsolicited disclosure.

Third party harm can also occur outside families, when genetic research is carried out on minority groups who share a similar genetic make-up. In this case it is quite possible that group members who did not take part in research will be affected by its outcome (indeed, those who do participate might be affected in unanticipated ways). For instance, susceptibilities for certain conditions may be found – thus 'particular mutations predisposing to breast, ovarian, and colon cancer have been identified through studies of Ashkenazi Jews' (Weijer and Emanuel, 2000, p.1142). It is certainly conceivable that the community might be discriminated against on this knowledge, for instance by health insurance providers. In this case, research outcomes are disadvantageous for a whole group and not only for those who take part in research. It would seem that this slight but alarming possible effect of genetic research must also be conveyed during the informed consent process.

The active governance of human genetic banks

As we have seen, human genetic banking is increasingly taking the form of large sample collections, where donors (or their parents) have given broad, though probably not unlimited, consent for future research. Concerns about the idea of informed consent as the sufficient legitimating ground of future use of these samples and associated data are not difficult to envisage.

As the previous section has stressed, the information involved is complicated, especially with regard to possible future uses and benefits. One might expect donors to grasp that use of their samples and health data may help generate various possible benefits, with remote but complicated risks of harm to them. But they will neither be allowed, nor arguably capable of, any precise input into the research that will be conducted. (At most, they may be able to 'opt out', if they disapprove of the way research initiatives are heading;

because of informational constraints, among other factors, one would expect this to be done only by a very few donors.) Never mind the limits of most persons' grasp of the research science and the political economies of commercially driven medical research; researchers themselves will only have a tentative grasp of some of the possibilities for future investigation, much less a thorough understanding of the justice or injustices of pharmaceutical companies' interaction with national health systems. Consent may be there, but its basis in 'information' will, despite all best efforts, be relatively slight. There is, moreover, very little prospect of short- or medium-term benefit to participants. In other words: sample collections are established on the good will and altruism of individual donors, and donors must invest a significant degree of *trust* – in those who will manage the database, and with regard to the uses to which it is put.

Even if one is less sceptical than we are, about how 'informed' consent can be in these cases, clearly it can only play a limited role in legitimating such projects. Suppose, for example, enough donors were to agree to a collection and corresponding research with worrying eugenicist overtones. Just because enough people provide their informed consent, this is hardly enough to legitimate a collection and its planned uses. Correspondingly, a collection that is badly underused or which, however unintentionally, becomes the dominant preserve of a few pharmaceutical companies, would surely represent a questionable enterprise, not least given the altruistic origin of the data bank.

This consideration only sharpens the question of how far gene banks will really deliver benefits (cf. Chadwick, 1999). The focus on informed consent has, we believe, led to inadequate attention toward those to whom trust is being given, and what would *vindicate* this trust. Pharmaceutical companies are provided with an extremely valuable resource for which none of the donors will be financially compensated. This altruism is surely matched by a moral responsibility to use the resource, at least in part, for the common good.

Bearing in mind this problem of public benefits, Berg and Chadwick (2001, p.320) suggest that we should rethink some of the attitudes that surround informed consent, and that an alternative ethical approach might be found in the political notions of solidarity and equity. They believe that it may be appropriate to question the right of research subjects to withdraw from these banking projects, once they have consented to inclusion in the first place, arguing that we have duties of solidarity to contribute to medical research. The suggestion that people may have a duty to participate in research is, of course, not new and, especially where the risks to subjects are rather slight, quite reasonable – especially where health care systems embody strong elements of solidarity, to ensure that all citizens will benefit from the research, should be need arise. Problems emerge, however, if we think of such a duty as enforceable, given that there is considerable scope for reasonable disagreement as to what constitutes desirable research (or, indeed, acceptable risk). In the case of human genetic banking, the fact that consent cannot be given on the basis of full information, only in terms of the broad aims of future research, means that withdrawal is the only effective way of preserving *any* sort of choice for individual data subjects as to the fate of their samples and health care data.

Moreover, as the emotive example of Alder Hey has surely demonstrated, informed consent has become a crucial part of the "political settlement" between researchers and the public. If it were explained to potential research subjects that their one-off consent, based on necessarily partial information about future research possibilities, were final, this would surely be damaging in terms of recruitment and, quite possibly, negative publicity.

In contrast, we believe that informed consent, with the on-going possibility of withdrawal, should not be *supplanted* by alternative approaches, but rather be *supplemented.* Whether we think of solidarity and equity, or the trust that donors are investing alongside their samples, it seems to us that the priority must be *to ensure that genetic banks are used for worthy, publicly endorsed ends*. To get a practical grasp on this, we do not need to limit the rights of research subjects, any more than we need insist on an unrealistic level of understanding in the consent process. Instead, we must focus on the duties of those who create and manage genetic banks – duties to ensure their exploitation for real public benefit or, in other words, duties to make notions such as solidarity into working realities.

Kaye and Martin (2000) have rightly drawn attention to a crucial difference between the Icelandic database and the MRC/Wellcome proposal. Though the Icelandic proposal seems much more commercially oriented, and considerably less 'voluntary' in its inclusion of samples and data, it at least was the subject of vigorous public and political debate. On the other hand, BioBank UK, certainly it its earlier stages, seemed to be proposed with little view to public consultation or independent oversight of the collection's use. Nor, one might add, can opinion sampling (Porter, 2000; People, Science and Policy, 2002) form an adequate substitute for political debate on the desirability and aims of such a genetic bank. We suggest that one important focus, and a possible conclusion, of this debate should concern *the public responsibilities involved in the governance of large sample collections.*

In the first place, there are a series of 'negative responsibilities', which all contributors take for granted. These include a duty to ensure that proper procedures are followed in gathering samples and information – one hopes this may be taken as read, after recent organ storage scandals. There is also a duty to ensure the security of confidential data – which may be a more challenging task, especially given the cavalier attitude to privacy rights shown in the development of the national police database (Bingham, 2001) and, one might argue, in the police seizure of material from the Scottish HIV study, discussed above.

Further, there is an uncontroversial duty to review the ethics of research proposals, via research ethics committees. Such bodies should veto, or help reformulate, research proposals that are unnecessary, scientifically unsound, or ethically undesirable. While it is obvious that research using human genetic banks should be subject to this scrutiny, the problem is that such review is essentially negative: it can only stop unethical or unwanted research from being undertaken, but cannot pro-actively steer the usage of a DNA bank. Such committees are almost wholly dependent on the proposals that come before them, and are in no position to satisfy donors' trust that genetic banks will be well-used and maximally contribute to the public good.

These essentially protective roles are likewise alluded to in the so-far brief comments about the governance of BioBank UK (People,

Science and Policy, 2002, p.27f; but cf Wellcome Trust, 2002, p.2.3.1). They do not, however, capture what we see as the main concern: structures of research governance that will fulfil a positive duty to ensure that worthwhile research is conducted using the collections. We see such structures, alongside the institutions which fund or support large sample collections, as having a much more onerous *positive* duty of *soliciting and prioritising research that is likely to realise the ends set by public and political debate.* In the case of wealthy grant-giving institutions such as the Wellcome Trust and the MRC, moreover, there seems to be a clear imperative to *enable* such research by funding.

It is here that the limitations of the Icelandic bank are most obvious. While the founding legislation speaks of 'a duty to use the data' (quoted and critically discussed by Rose, 2001, p.17), and while the licensee is indeed in a position to 'use the data' fairly extensively, the problem is that this use will be determined by the licensee only, that is, by a private company. Even if this is done within adequate protective and research-ethical safeguards, the research will be primarily driven by commercial imperatives, which have only uncertain and contingent relation to the public good or public priorities. The North Cumbria initiative, by contrast, certainly does not face this problem. Its difficulty, instead, is that it is in no position to enact public research priorities. Though it was created only after extensive public consultation, and carefully publicises the research undertaken using the bank, if proposals are not made, or funding not forthcoming, then nothing can be done.[6]

BioBank UK is, potentially, in a much more fortunate situation. The founding partners are extremely powerful in determining research priorities in the UK, being among the relatively few non-profit organisations in the world which compare in spending power with the international pharmaceutical companies. These partners face, we suggest, imperatives not only to publicise research projects and future priorities, but also to be open to public debate regarding these. It may be, for example, that custodians of BioBank UK find that there is pressure for coordinated, systematic research on common, chronic conditions which have attracted only patchy research – perhaps because they are rarely fatal, although they may cause widespread misery (for example, asthma, hay-fever or irritable bowel syndrome).[7] It may be also, as those examples suggest, that there is an imperative for research that does *not* focus primarily on (potentially) very profitable drug treatments. That is to say, the sponsors and guardians of such large databases will have a duty to focus resources and researchers' attention on areas where pharmaceutical companies have little or no incentive to invest. At the same time, there will be a clear need to monitor and report on commercial exploitation of sample collections. As the Human Genome Organisation Ethics Committee (2000) has suggested, this may involve seeking, and reviewing, mechanisms of benefit-sharing. Not least, there will be a need to provide regular public reporting on *actual* benefits obtained from research upon the collections, and not only the speculated benefits so often emphasised in these debates (cf. also O'Neill, 1999).

Conclusion

Human genetic banking for pharmacogenomics or population genetics is likely to increase significantly in the near future. Informed consent procedures need to be adapted for use in human genetic banking as this activity carries a different set of often complicated potentials for harm than 'traditional', non-genetic research. Particularly issues of feedback, future uses (including commercial exploitation), confidentiality, and possible harm to third parties should be communicated to donors.

This is already a demanding list, but it still necessarily involves partial information, given the limited degree to which researchers can foresee possible uses: not only are possible harms complicated and uncertain – so too are possible benefits. We have therefore explored the supplementary notion of duties of custodians and public sponsors, to ensure that samples are used for the public good and for publicly endorsed ends. Given the predominance of commercial research interests, and the often haphazard nature of coordination of researchers' efforts, the grave danger exists that sample collections will be selectively exploited on terms set by commercial partners. As the banks rely on donor altruism for their existence, such exploitation has to be regarded as inequitable and unacceptable. Even if some benefits were to emerge, this would hardly represent systematic exploitation of the collection for the benefit of the whole community. Research subjects should not have to consent to less than this, and – we have argued – sponsoring institutions and custodians have a duty to ensure that donors, in fact, do not.

Notes

1. These are currently standard examples; they are also discussed, for instance, in Berg and Chadwick (2001).
2. For the problematic nature of this, including doubts as to whether opting out will be a genuine option for any but a very small number of adults in the first years of the study, see Rose (2001, p. 24f).
3. Data may be classified in terms of the 'degree' of anonymity: 1) *Anonymous* – no information about the donor available when the data were filed. 2) *Anonymised* – information about donors destroyed. 3) *Encoded* – information about donors coded with a serial number and the key held elsewhere. 4) *Encrypted* – data turned into meaningless strings for commercial security. 5) *Identified* – information about donor kept with the data, without coding (Spallone and Wilkie, 2000, p. 199; European Society of Human Genetics, 2001).
4. Various methods may be used to increase data and sample security, for instance, as in North Cumbria, by storing samples and coding information at separate sites (North Cumbria Genetics Project, 2000, p. 8).
5. Though we think this 'public interest' claim extremely dubious in the case at hand, the point here is only that such cases do occur, and may do so legitimately.
6. Since the collection is on nothing like the same scale as the other two, representing a much smaller investment of public resources and overall donor effort, we stress that we do not mean this as a criticism of the North Cumbria project or other relatively small initiatives.
7. Cf the list of research hypotheses mentioned in the draft protocol, Wellcome Trust, 2002, p. 2.1.4.

References

AstraZeneca (2000) 'Memorandum – Evidence to the Select Committee on Science and Technology' www.publications.parliament.uk/pa/ld199900/ldselect/ldsctech/115/115we07.htm

Bell, J. (2000) 'Memorandum – Evidence to the Select Committee on Science and Technology' www.publications.parliament.uk/pa/ld199900/ldselect/ldsctech/115/115we50.htm

Berg, K. and Chadwick, R. (2001) 'Solidarity and Equity: New Ethical Frameworks for Genetic Databases', *Nature Reviews*, 2, pp. 318-321.

Bingham, R. (2001) 'A Database of the Innocent?', *Splice*, 7 (2/3), pp. 8-9.

Brody, B.A. (2001) 'A Historical Introduction to the Requirement of Obtaining Consent from Research Participants', in L. Doyal and J.S. Tobias (eds) *Informed Consent in Medical Research*, BMJ Books: London, pp. 7-14.

Brown, A.L. (2001) 'A Confidential Con Job', *Times Higher Education Supplement*, 1489 (1 June 2001).

Chadwick, R. (1999) 'The Icelandic Database – Do Modern Times Need Modern Sagas?', *British Medical Journal*, 319, pp. 441-444.

Chadwick, R. (2001) 'Informed Consent and Genetic Research', in L. Doyal and J.S. Tobias (eds), *Informed Consent in Medical Research*, BMJ Books: London, pp. 203-210.

Clarke, A. (2001) 'Genetic Counselling', in R. Chadwick (ed.) *Ethics of New Technologies*, Academic Press: San Diego, pp. 131-146.

Clarke, A., English, V., Harris, H. and Wells, F. (2001) 'Ethical Considerations', *International Journal of Pharmaceutical Medicine*, 15, pp. 89-94.

Department of Health (2000) 'Memorandum – Evidence to the Select Committee on Science and Technology'. www.publications.parliament.uk/pa/ld199900/ldselect/ldsctech/115/115we17.htm

Department of Health (2001) 'Milburn Promises Reforms After Alder Hey Inquiry and Pays Tribute to Parents', Department of Health, News Desk. www.doh.gov.uk/newsdesk/latest/4-naa-30012001.htm

Dyer, C. (2001) 'Use of Confidential Data Helps Convict Former Prisoner', in *British Medical Journal*, 322, p.633.

Dyer, C. (2001a) 'Scientists fear breach of confidentiality will threaten research'. www.guardian.co.uk/Archive/Article/0,4273,4153840,00.htm

European Society of Human Genetics (Public and Professional Policy Committee) (2001) 'Data Storage and DNA Banking for Biomedical Research: Proposed Recommendations' (draft consultation document), www.eshg.org

Glaxo Wellcome (2000) 'Memorandum – Evidence to the Select Committee on Science and Technology'. www.publications.parliament.uk/pa/ld199900/ldselect/ldsctech/115/115we24.htm

Haraldsdottir, R. (2000) 'Fire and Fury in Iceland', *Science and Public Affairs*, February 2000, pp. 12-13.

House of Lords (Select Committee on Science and Technology) (2001) 'Human Genetic Databases: Challenges and Opportunities'. www.publications.parliament.uk/pa/ld199900/ldselect/ldsctech/57/5701.htm

Human Genome Organisation (HUGO) Ethics Committee (2000) *Statement on Benefit-Sharing*, Human Genome Organisation, London.

Kaye, J. and Martin, P. (2000) 'Safeguards for Research Using Large Scale DNA Collections', *British Medical Journal*, 321, pp. 1146-1149.

King, D. (2000) 'Memorandum – Evidence to the Select Committee on Science and Technology'. www.publications.parliament.uk/pa/ld199900/ldselect/ldsctech/115/115we23.htm

Lowrance, W.W. (2001) 'The Promise of Human Genetic Databases', *British Medical Journal*, 322, pp. 1009-1010.

Martin, P. (2000) 'Memorandum – Evidence to the Select Committee on Science and Technology'. www.publications.parliament.uk/pa/ld199900/ldselect/ldsctech/115/115we52.htm

Medical Research Council (2000) 'Memorandum – Evidence to the Select Committee on Science and Technology'. www.publications.parliament.uk/pa/ld199900/ldselect/ldsctech/115/115we32.htm

Moreno, J.D., Caplan, A.L. and Wolpe, P.R. (1998) 'Informed Consent', in R. Chadwick (ed.) *Encyclopaedia of Applied Ethics*, Academic Press, San Diego, pp. 687-697.

North Cumbria Community Genetics Project (2000) *Report 1996-2000*, Westlakes Research Institute.

North Cumbria Community Genetics Project (2001) *Further Information*, Westlakes Research Institute.

North Cumbria Community Genetics Project (2001a) *Informed Consent Form*, Westlakes Research Institute.

O'Neill, O. (1999) 'Genetic Information and Ignorance', Greenwall Lecture delivered to the American Society for Bioethics and the Humanities, unpublished.

People, Science and Policy Ltd. (2002), *BioBank UK: A Question of Trust: A consultation exploring and addressing questions of public trust*, Report prepared for the Medical Research Council and the Wellcome Trust, People, Science and Policy, London.

Porter, T. (2000) *Public Perceptions of the Collection of Human Biological Samples*, Report prepared for the Medical Research Council and the Wellcome Trust, Cragg Ross Dawson, London.

Rose, H. (2001) *The Commodification of Bioinformation: The Icelandic Health Sector Database*, Wellcome Trust, London.

The Royal Liverpool Children's Inquiry (2001) 'Summary and Recommendations', www.rlcinquiry.org.uk

Spallone, P. and Wilkie, T. (2000) 'The Research Agenda in Pharmacogenomics and Biological Sample Collections', *New Genetics and Society*, 19, pp. 193-205.

Weijer, C. and Emanuel, E.J. (2000) 'Protecting Communities in Biomedical Research', *Science*, 289, pp. 1142-1144.

Wellcome News (2000) 'A Sample Solution', in *Q3 – Wellcome News*, 24, pp. 10-11.

Wellcome Trust (2002) *Draft Protocol for Biobank UK (February 2002)*, www.welcome.ac.uk/en/1/biovenpopprt.html

Wolf, R. (2000) quoted at: BBC News Online 'Made-to-measure-medicine', news.bbc.co.uk/hi/english/health/newsid_704000/704577.stm

Womack, C. and Gray, N.M. (2000) 'Human Research Tissue Banks in the UK National Health Service: Laws, Ethics, Controls and Constraints', *British Journal of Biomedical Science*, 57, pp. 250-253.

Yirrell, D.L., Robertson, P., Goldberg, D.J., McMenamin, J., Cameron, S. and Leigh Brown, A.J. (1997) 'Molecular investigation into outbreak of HIV in a Scottish prison', *British Medical Journal*, 314, pp. 1146-50.

9 Regulation and Social Perceptions of Genetic Data Banking in Germany

Juergen Simon and Susanne Braun

Introduction

In the last few years there was an important expansion of human DNA sampling and data collecting in order to exploit and study the genetic information collected. In the next years the storage and use of such genetic information will be of an increasing importance. Actually the German government discusses the use of genetic data in Labour Law, research and for insurance purposes.

As in other countries a lot of databanks already exist in Germany and the genetic testing commerce has become an important marked with a global total turnover of US $1.3bn.[1] The potential benefits seems to justify the establishment of genetic databanks but the possibility of misuse imposes a responsibility of proper management and protection of the subjects' interests. The availability of personal genetic information poses many problems concerning privacy, confidentiality of the data and informed consent, because genetic data are highly specific information, revealing facts not only about the examined person but about the members of his or her family and having therefore a great impact on a person's life or lifestyle. Therefore genetic research has to be conducted with sufficient safeguards to protect individual interests, without obstructing legitimate medical research activities of benefit to society. New forms of discrimination have to be avoided when insurance companies or employers could use genetic data as a reason for denying insurance cover or turning down a person for a job. The European Parliament stated:

> The use of and access to personal genetic information should be debated with a view to legislation, which should particularly focus on protecting the

> individual's personal integrity on the requirement to obtain his consent... Member States should protect individuals' right to genetic confidentiality to ensure that genetic profiling is used for purposes beneficial to individual patients and society as a whole; there should be an exception to this general principle of confidentiality where genetic fingerprints held in DNA databases are used to identify and convict criminals.[2]

In the following, the different reasons for genetic data banking, the social perception and the quality management of databanks will be discussed. Then the legal aspects concerning genetic data banking in Germany will be examined.

Reasons for genetic data banking and status of collections

There is a wide range of application fields of genetic data banking. Above all, the medical sector is the most important area. The origin of diseases could be detected and new diagnosis and therapy methods could be more efficiently developed by constructing genetic profiles. The genetic data registration of entire populations or groups makes the construction of genetic profiles possible. Furthermore the use of genetic fingerprinting in criminal cases, on the legal basis of § 81 a, c and completed by § 81 e, f, g Criminal Procedure Code, has been operationalised and developed through court rulings. Finally genetic data could be used for certification of parentage. A clear distinction exists between legislation and policy that relates to criminal databanks and that which relates to medical databanks. The further remarks will focus on the medical sector.

Concerning the status of collections several types have to be distinguished. In anonymous collections the biological materials were originally collected without identifiers and are impossible to link with their sources. In anonymized collections, biological materials were originally identified, but have been irreversibly stripped of all identifiers and are impossible to link to their sources too. In identifiable collections, biological materials are unidentified for research purposes, but can be linked to their sources through the use of a code. In identified collections, identifiers, such as a name, patient number, or a clear pedigree location, are attached to the biological materials.

Social perception of genetic data banking in Germany

The German Ministry for Education and Research states in its report upon the human genome research in Germany[3] that the genetic epidemiology is still underdeveloped without starting any public debate. For 30 years, population screening has been available in Germany. These screenings were offered and can be voluntarily done. Screening is a kind of test performed for the systematic early detection or exclusion of a hereditary disease, the predisposition to such a disease or to determine whether a person carries a predisposition which may produce a hereditary disease in offspring. This genetic screening has to be distinguished from genetic testing because the implications are different and it does not necessarily lead to the prevention or treatment of diseases, whereas genetic testing is carried out on patients, who for whatever reason have taken the initiative and seek advice. But it is only the beginning of population studies in Germany. For instance in the databank of the Berlin enterprise InGene data of more than 3000 voluntaries have been registered since the first of April 2001. This number should increase up to 40.000 per year.[4] Seven hospitals and about twenty physicians collect data, resulting of a blood test, medical data and a large questionnaire of 23 pages. Clinical history, living habits, social surroundings and environmental conditions have to be respected too.

For InGene the human phenotype is the starting point of research. On this base populations and later on gene profiles should be established. The detection of new disease relevant genes should be possible. This "direct and accelerated revelation of clinical relevant genetic factors" will be important for further medical development"[5] and a contribution to a future personalised medicine. The necessary votes of the ethic committees exist,[6] authorising InGene to transmit data and sample to third persons (this means DNA and serum sample as well as the above mentioned clinical questionnaires).

With this project private enterprises try to realise a small version of what has been funded by private organisations or government in Iceland or with the governmental supported gene databank in Estonia. The Estonian example of a central health databank, supported by a commercial exploitation enterprise, is widely accepted on the international level. Personalised medicine should better contribute to the individual provision and insofar economize costs. Therefore a working group at the World Medical Association is actually elaborating ethical guidelines organising the development and function of genetic databanks. Especially these guidelines will be made for governments planning national gene data banking.

In Germany great projects as in Estonia would not be accepted and are not planned[7] because there is no social consensus for this kind of projects. It is mostly understood as an eugenic tendency and related to fears of the transparent human being. The possibilities of electronic data processing associated with the fear of a total transparency of the citizen impedes extensive governmental data storage and data exploitation in Germany.[8] Since the beginning of the eighties legislation, jurisdiction and science have intensively discussed this subject under data protection aspects with the result that the respect of the fundamental rights is the essential leading principle of the whole bio politics and particularly for genetic data banking.

Management, quality control and security issues

Many DNA banks are concerned with how to obtain valid informed consent, safeguard the privacy of samples and data and avoid potential misunderstandings with depositors. The value of a collection is proportional to the amount and quality of the information attached to it. The full benefits for which the subjects gave their samples will be realised through maximizing collaborative high quality research but the multiplicity of actors and of rules that regulate them (public versus private, hospitals, research centres, laboratories ...) make the situation very difficult to comprehend.[9] Rules for exchange and sharing of information should exist, but the status of collections often is not known and most laboratories have no written policies or agreements regarding this activity.

Because of the sensibility of the results security mechanisms to ensure the confidentiality and long-term conservation of genetic information is an absolute condition and has to be implemented for instance for the quality of genetic examinations (validity and exactness), the employees and the procedure of genetic examinations.[10] Genetic examinations should only be made by physicians and indications should be made by human geneticists or special physicians[11] to guarantee an adequate information, consultation and essential protection of concerned and third persons. There has to be a safeguard against unauthorised access to genetic data banking as well as the safeguard against the use of the databank for anything other than their overriding purposes.

Legal aspects of genetic data banking in the medical area

In Germany no special law of genetic data banking exists. Information can only be used according to the constitutional principles especially the fundamental rights and the data protection regulations.

The fundamental rights of concerned persons

Strong legal positions of the concerned persons, especially the donators of the information, are needed. The Constitutional situation depends on the kind of data banking system, if it is a governmental or a private one, because fundamental rights are only defensive rights against the State and cannot be directly applied between privates.[12]

First of all dignity, the highest value of the German Constitution (GG) and protected by the "guarantee of eternity", in Art 79 III GG, has to leave unimpeachable, Art. 1 I GG. Every human being has a dignity in the sense of a social claim of value and respect because of his being as a human.[13] The principle of human dignity is based on the idea of human beings focusing at the same time on the inherent value and a common relation element.[14] A legal argument as Art. 1 I GG is not necessary for these inalienable and inviolable rights, but responsible for the qualification of human dignity as a legal notion and open for legal interpretation.[15] The content has to be determined with view to Art. 1 I s. 2 GG, the obligation for all governmental authorities, to respect and protect the dignity. But it is difficult to explain the notion human dignity. Following a kind of negative definition, human dignity is affected, if the single human being is degraded to an object, to a mere mean, to a replaceable unit.[16] It is not compatible with human dignity to accept a treatment calling their subject quality in question.[17] The State is not only obliged to omit actions being an offence to human dignity but as well to act in a defensive way so to avoid offences and attacks to human dignity from third parties.[18] So all bio scientific developments will be accompanied by the question if they are compatible with human dignity or if there will be an offence to human dignity which has to be restricted by the State,[19] but avoiding a general emotional appeal to human dignity in form of a knock-down-argument.[20] The other constitutional provision have to be interpreted within the light of the human dignity, being at the same time a concretisation of the notion human dignity.[21] At the beginning of the discussion about genetic testing in Germany, Benda, the former president of the Federal Constitutional Court, supposed an intrusion of human dignity in case of a total storage of individual hereditary factors.[22] The storage of single genetic characteristics seems not to be an offence against human dignity, although associated with intensive effects for the concerned person.[23]

The use of personal data like the name of the patient or their symptoms related with a gene databank should not violate the general personality right, which has been derived by the Federal Constitutional Court from Art 2 I GG in connection with Art 1 I GG. The task of this right is to guarantee the closer personal sphere and its basic prerequisites.[24] Art 2 I GG guarantees expressively the general freedom of action, an active element, including the respect for the inner personality sphere, the intimate and private sphere as the preservation of their basic prerequisitions.[25] But Art 2 I GG could be restricted. Concerning the general personality right the Federal Constitutional Court developed the sphere theory as a marking point for these restrictions. The basic sphere, an inviolable domain of private life style, is absolutely protected and every action of executive organs will be prohibitied.[26] Beyond this sphere, the domain of private life style being in a social relation could be restricted if there is a predominant interest of the public and a strict respect of the principle of proportionality.[27]

These contents of the general personality right will be essential especially with regard to the modern biotechnology developments and the associated dangers for the personality and individuality.[28]

Therefore the general personality right will be an adequate criterion for a judgement about the legal aspects of gene data collection and storage in relation with the fundamental rights.[29] Even if the Court has not yet developed concrete criteria whether genetic data belong to this absolutely protected sphere, it has to be supposed that the decision, if the individual would like to know the genetic details of his future health, belongs to this inviolable essential content.[30]

A part of this general personality right is the right of informational self determination guaranteeing the individual person a free decision which personal data shall be given to whom, at which time and for which purposes these data can be used.[31] The person has to know about the probable consequences for herself when using these information.[32] This right is not limited to the automatic data processing.[33] Another part of the right of informational self determination is the right not to know,[34] especially of the family members of the examined person who do not want to know about their genetic constitution. It is not necessary to argue directly with the basic fundamental right of human dignity, because the general personality right is based on Art. 1 I and so directly related to the principle of human dignity.[35] For the individual it is important to have the freedom to develop their own personal identity without being burdened with a foreknowledge.[36] Consequently the governmental instruction concerning data storage would principally be forbidden[37] as well as any sanction in case of refused consent. An exception could be the protection of the right of life of third persons and the preservation of serious health damages.[38]

If the individual should know his or her hereditary factors or perhaps lethal or later appearing diseases against their will, there are convincing arguments to deny a global intrusion of human dignity but there will be an offence against the general personality right. So much the more if the disease could appear in the near future and then perhaps the curative ability of most human beings to suppress would be overstrained.[39] The typical uncertainty related with the storage of genetic data, if the stated disease risk will be realised, would be another burden.[40] Because the corresponding diagnosis could only be statistically-epedemiologcially interpreted for a group of persons. The individual has to live with the uncertainty. This could be a threat especially in case of lethal disease risks whose intensity could vary from individual to individual and even lead to an existential conflict.[41] Only a right not to know could avoid the probable loss of impartiality, frankness and finally freedom towards the own future. Meanwhile this meaning is widely accepted, generally recognizing the necessity of an informed consent.[42] The individual must have the possibility to choose the right not to know his hereditary factors even if healthy disadvantages are related with his decision,[43] even if the genetic data storage would only or mainly be made to discover endogenous health risks for preventive aims and to instruct later on preventive or therapeutically indicated participating obligations. The right not to know includes the free decision of the individual life style. It has to be at the individual's disposal, which concrete health diagnosis would threaten the individual so to ignore further information. A restriction of the right to know for instance in cases of severe diseases would not be compatible with the right of self determination. It would be different if the citizen defends his or her right not to know their hereditary factors against the interest of the State or other private persons or if he voluntarily offers material or information about his genetic constitution. This would be an expression of his personality. A prohibition to inform about his own genetic data, would contradict the right of life, physical integrity or free personality development, if it aims at a defence of disease risks or supporting research.

Finally, the right of the person whose genetic data are collected and stored could collide with the freedom of research protected in Art. 5 III GG. There are no particular reservations or limitations mentioned in Art. 5 III GG, but the so-called "constitutional imminent barriers"[44] as other constitutional values have to be respected and the colliding values have to be assessed.

Collision of the right to know and the right not to know

In case of an increasing correlation between genetic characteristics of an individual and certain diseases, more and more negative stigmatising social prejudices towards the concerned person or groups of persons will be expected. In principle everyone has the right to keep secret their genetic diagnosis instead of revealing the genetic diagnosis, insofar a situation demanding for an intensive protection exists. But there will be problems, if in case of a genetic diagnosis the right to know of the examined person collides with the right not to know of the same persons or another person,[45] for instance a family member. If a person is positively tested for Huntingdon's Chorea and her grandfather already had this disease, then it will be sure that one parent would be carrier of this disposition as a connecting link and would get this disease. This is a constellation of private family relations which cannot be solved with regular legal measures. The prohibition not to disseminate the test result or not to tell genetic data to any person cannot effectively be established inside a family community.[46] One of the results of an international study about the handling of genetic data with 1400 patients from USA, Canada, France and Germany was, that only 500 persons demanded a right not to know, and many patients wanted information for the whole family even if single members do not agree.[47]

It has to be asked if the right not to know can be guaranteed even in such a constellation. So among others a restrictive access to genetic examinations could be possible. But this would mean that genetic testing is not available for anyone. The contradictory legal positions and interests have to be assessed and criteria have to be developed, who and under which conditions could make genetic testing.[48] The right to know would be more important if the concerned subject would be of a higher value, for instance if a testing result would be of great influence for life styling whether by a therapeutic treatment or a prophylactic life style of the concerned person. At the same time the other person has so much more the right not to know. A result free of contradiction would not be possible. Anyhow, the necessary genetic testing must be accessible for those persons, whose serious disease could be efficiently treated.[49] But the concrete criteria for this access in the individual cases doesn't exist and it seems very difficult to define them.

Regulations of the Federal Data Protection Law

The collecting, storage and use of genetic information has to be done according to the regulations of the Federal Data Protection Law[50] (BDSG). This means on the one hand the law protects the individual and their right of informational self determination and on the other hand it is possible to control the databanks by this law.

The Federal Data Protection Law is very complex, since the applicable law depends on the status of the data collecting and storing institution (public, private, federal, state), it contains different permissions for collecting, storing, using, transferring for own and other purposes and there are numerous exceptions in other laws. With respect to the principles of § 3 a BDSG to collect personal data only if it is really necessary and to use anonymous data if possible, the special purpose of this law is the protection of the individual, so that the use of his data would not violate his general personality right (§ 1 I BDSG).[51] Data are single facts about personal or material relations of an identified or identifiable person. Single facts in turn are information about the physical or mental situation, which can be discovered by genome analysis. Data procession and uses are only permissible to the extent that is authorised by the law or another legal provision or if the concerned people have consented (§ 4 I BDSG). This consent has to result of the free decision on the base of an intensive information about the intended purpose of the collecting, storage and use (§ 4 a BDSG).

The law does not relate specifically to the protection of personal genetic data, nevertheless the general regulations may apply to the collection, storage and use of personal genetic data qualifying them as personal data in the sense of (§ 3 IX BDSG).[52] § 28 III No 1, 2 BDSG allows the use and transfer for danger preventive or criminal procedure purposes, despite the initial purpose.

Referring to the second aspects of the BDSG, the controlling of databanks, it has to be stated that controlling authorities exist to supervise the respect of data protection regulations (§ 38 VI BDSG). In case of defiance of the rules, the authorities can sanction the databank.[53] Furthermore non-public databanks in the sense of § 2 IV BDSG are obliged to indicate their activity starting to the controlling authority (§ 4 d BDSG). If they forget this indication, they have to pay a fine (§ 43 I No 1, 2, III BDSG). Insofar the Federal Data

Protection Law establishes as well a certain control mechanism for databanks.

The Data Protection Law of the States include all public authorities of the respective states. They have to be applied in public hospitals or the state or the municipalities. The provisions are subsidiary to special provisions in other laws. In most state hospitals laws exist containing special data protection rules.

Results

The survey and the reflections about the social perception and the legal aspects of genetic data banking in the medical sector showed that above all the main problems focus on the confidentiality, privacy, discrimination. Every breach of confidentiality could impact insurability, destroy family relationships and causes stigmatisation or discrimination. Therefore transparency of the activities of genetic data banking is necessary for the individual related with intensive information (existence, ownership, application and group of persons covered by the databank) maintaining at the same time a status of anonymity concerning the concrete data. It is obvious that the sampling, storage and examination of body material without knowledge and consent of the concerned person would be an offence against the right of informational self determination. Even the already formulated consent declarations often are not enough determined or temporarily limited. Whether it will be necessary to create documentation and reporting obligations disregarding the principle of appropriation has to be discussed. But it must be clear for the concerned person to know who, how far, for which purpose and where his genetic data will be used. Additionally the person must be able to revoke the consent.

Furthermore a data protection level and system has to be established in Germany corresponding to the particularity of genetic data transforming the informational self determination right and preserving at the same time the right not to know,[54] especially because the right to data protection has finally been recognised as a fundamental human right in Art. 8 of the EU-Charter of Fundamental Rights and because these prequisitions are not realised by the existing Federal Data Protection Law.[55] The regulations of this law contain more or less general clauses and undetermined legal notions needing an interpretation for each single case. The Data Protection directive of the EU[56] established new measures concerning the handling of patients' data in medicine and health care, forbidding the collection of data about health and sexual life (Art. 8 I). Those data are protected. Art. 8 III foresees several exceptions if the data are really necessary and an obligation to keep the identity of the concerned person a secret. These exceptions have to be interpreted when transferring the directive into national law. Until now such a transfer into German law doesn't exist. In 1999 the central ethics committee of the German Medical Association gave a statement concerning the use of patients data in medical research and in health policy.[57] Informed consent will be necessary requiring the individual being provided with information concerning the purpose of the research, whether information obtained is to be coded, deidentified or identifiable, privacy protections, that the result of the research may be commercialised, whether samples will be stored for future research purposes and that the consent may be withdrawn at any time. The committee demanded for a better protection of the general personality right concerning the use of patient data for medical research. Ethical and legal problems arise if the concerned person cannot consent and consent cannot be sought from the person's legal representative. Then an ethical justification will be necessary. But if a great number of data should be collected, it will be too difficult and expensive to ask each patient for his consent, if the storage or use will be made for another than the initial purpose, or data should be collected from different origins, or information about a patient should be permanently collected or data should be stored for a long time. In these cases the interests of both sides, the databank and the patient or other concerned persons have to be intensively assessed.[58] A total prohibition of collection, storage and use of personal data certainly cannot be justified because essential functions of health policy would be destroyed especially if the citizen is legally obliged to finance the health system. In the contrary any derestriction in the data protection field could be a risk for the general personality right. This problematic and difficult situation can only be handled if concrete legal measures will be defined, incidentally contributing to a better social perception of genetic data banking. Actually, in Germany a draft Genetic Testing Law exists, containing special data protection regulation: "Genetic data should only be collected, stored or used for the purpose of genetic examination and only so far as the collection, storage or use corresponds to the informed consent of the concerned person".[59] This draft could be the first step towards a new special regulation concerning genetic data banking.

Notes

1 Goerdeler/Laubach, ZRP 2002, 115 (116).
2 European Parliament resolution A4-0080/2001 on the future of the biotechnology industry.
3 BMBF (2001), p. 15.
4 Berliner Zeitung vom 28.12.2001.
5 www.ingene.de/plat_tech.html.2501.02.
6 Ethikkomission der Universität Witten-Herdeke. Datenschutzkommission Berlin und Ethikkommission der Landesärztekammer Berlin.
7 Schwägerl, Gleiche Gene, Deutsch-estnische Biopolitik in der Charite, Frankfurter Allgemeine Zeitung, 21.1.2002.
8 Simon/Taeger in 1981.
9 European Society of Human Genetics Public and Professional Policy Committee, background document, 30.10.2000, p.3.
10 12 See especially the directive 98/79/EG about in-vitro-diagnostic devices, L 331, p.1. Actually the German medical product law, which could be applicable concerning the used instruments, is being revised because of this directive, see BT-Dr. 14/6281.
11 Goerdeler/Laubach, ZRP 2002, 115 (117).
12 BverfGE 21, 369; 50, 336f. 68, 205.
13 BverfGE 87, 209 (228); Schmidt-Bleibtreu/Klein (1999), Art. 1 Rn. 1a.
14 BverfGE 4, 7 (15).
15 V. Műnch/Kunig (1992), Art. 1 Rn. 18.
16 Schmidt-Bleibtreu/Kelin (1999), Art. 1 Rn. 12.
17 BverfGE 27, 1 (16); 30, 1 (26).
18 See V. Műnch/Kunig (1992), Art. 1, Rn. 29 f; BverfGE 1, 104.
19 See Simon (2000), p.230.
20 Petermann (1996), p. 123: "The inflationary use of the human dignity argument has not always led to greater clarity".
21 BverfGE 6, 36; Schmidt-Bleibtreu/Klein (1999), Art. 1, Rn. 1.
22 Benda (1985), p.33: Totalsequenzierung.

23 BverfGE, NJW 2001, 879: The Federal Constitutional Court decided that is not possible to construct a personality profile when examining the non-coded part of DNA.
24 BverfGE 54, 148 (153).
25 BverfGE 54, 148 (151); Gretter (1994), p.26.
26 BverfGE 27, 1 (6); 27, 344 (350).
27 BverfGE 6, 389 (433); 27, 344.
28 BverfGE, NJW 1980, 207Off.
29 BverfGE 54, 148 (153); 65, 1 (141).
30 Donner/Simon (1992), p. 17; Rademacher (1989), p.736.
31 BverfGE 65, 1 (63).
32 Sokol, NJW 2002, 1767 (1768).
33 BverfGE 78, 77 (64).
34 Wiese (1991), p. 475; Donner/Simon, DőV 1990, 907 (912).
35 BverfGE 27, 344 (351). Its respect in legal relations with private third persons will be guaranteed by § 823 I Civil Law, foreseeing damages in case of violation of the general personality right.
36 Cramer (1991), p.255.
37 Rademacher (1989), p.736; other meaning Deutsch (1986), p. 1-4.
38 Wiese (1992), p.658.
39 See the second report of the "Interministerielle Kommission des Landes Rheinland Pfalz" 2989, p.45.
40 Bundesärztekammer, Richtlinien zur Diagnostik der genetischen Disposition für Krebserkennung, in: Deutsches Ärzteblatt 95 (22) 1995, A-1396.
41 Vitzthum (1991), p.69.
42 Donner/Simon (1992), p.5 ff, 14, 18; Schöffski (2000), p.121.
43 Donner/Simon (1990), p. 912 f.
44 BverfGE 28, 243 9162); 32, 98 (108).
45 Schneider: Wissen ist Ohnmacht, NZZ Folio, 09.2000 (Gene-Der Memsch und sein Erbe). http://www-x.nzz.ch/folio/archiv/2000/09/cover.html: Nancy Wexler, a researcher from a Chorea Huntington family said that she wants to know that she has not this disease, but she doesn't want to know that she has the disease.
46 Goerdeler, Gen-ethischer informationsdienst, Nr. 150, Febr./March 2002, p. 15.
47 Wertz/Nippert/Wolff/Aymé: Ethik und Genetik aus der Patientenperspektive: Ergebnisse einer internationalen Studie. Genomexpress 2/01.
48 Goerdeler, Gen-ethischer Informationsdienst, Nr. 150, Febr./March 2002, p.15.
49 Goerdeler, Gen-ethischer Informationsdienst, Nr. 150, Febr./March 2002, p.16.
50 The law was amended in May 2001, BGBI. I 2001, 904.
51 See Ordemann/Schomerus (1992), § 1 p.54.
52 Goerdeler/Laubach, ZRP 2002, 115 (117).
53 Measures or sanctions of trade law can additionally be taken, § 38 VII BDSG.

54 So the resolution of the federal and state commissioners for data protection during their 62nd conference from 24. –26.10.2001.
55 Goerdeler/Laubach, ZRP 2002, 15 (116).
56 Directive 95/46/EG from 24.10.1995, L281, p. 21ff.
57 Stellungnahme der Zentralen Ethikkommission bei der Bundesärztekammer, zur Verwendung von patientenbezogenen informationen für die Forschung in der Medizin und im Gesundheitswesen, in: Deutsches Ärzteblatt 96 (1999), A 3201-3204.
58 Stellungnahme der Zentralen Ethikkommission bei der Bundesärztekammer, zur Verwendung von patientenbezogenen informationen für die Forschung in der Medizin und im Gesundheitswesen, in: Deutsches Ärzteblatt 96 (1999), A 3201-3204.
59 See Art. 16 of the "Entwurf eines Gesetzes zur Regelung von Analysen des menschlichen Erbguts" (Gentest-Gesetz); http://www.gruene-fraktion.de/uthem/bildung/index.htm.

References

Benda, E. Erprobung der Menschenwürde am Beispiel Humangenetik. In: Aust Politik und Zeitgeschichte, Beilage zur Wochenzeitung "Das Parlament", B 3/85, 1985, S.33.

Bundesministerium für Bildung und Forschung (BMBF), Die Humanmgenomforschung in Deutschland, Bonn, 2001.

Bundesärztekammer, Richtlinien zur Diagnostik der genetischen Disposition für Krebserkennung. In: Deutsches Ärzteblatt 95 (22) 1995, A-1396.

Cramer, S. Genom- und Genanalyse – rechtliche Implikationen einer, Prädiktiven Medizin", Frankfurt, Bern, New York, Paris, 1991.

Deutsch, E. Die Genomanalyse. Neue Rechtsprobleme. In: ZRP 1986, 1ff.

Donner, H. and Simon, J. Genomanalyse und Verfassung. In: DöV 1990, 907ff.

Donner, H. and Simon J. Genomanalyse und Verfassung. In: Recht der Biotechnologie, Bd. III; Schwerpunktbeiträge, 2. Ergänzungslieferung, 1993.

Goerdeler J/Laubach B. Im Datendschungel. In: ZRP 2002, 115 (116).

Gretter, B. Gesetzlich geregelte Informationspflicht gegenüber risikoträgern von genetisch bedingten heilbaren Krankheiten? In: ZRP 1994, 24.

Münch, I. and Kunig, P. Grundgesetz-Kommentar, Bd. 1, München, 4. Auflage, 1992.

Ordemann, H-J. and Schomerus, R. Bundesdatenschutzgesetz, München, 5. Auflage 1992.

Petermann, T. Human dignity and genetic tests. In: Bayertz (ed.), Sanctity of life and human dignity, Dordrecht, 1996, p.123.

Rademcher, C. Zulässigkeit der Genanalyse? In: NJW 1989, 735.

Schmidt-Bleibtreu, B. and Klein, F. Kommentar zum Grundgesetz, Neuwied, 9. Auflage, 1999.

Schöffski, O. Gendiagnostik: Versicherung und Gesundheitswesen, Karlsruhe, 2000.

Simon, J. Die Menschenwürde als regulierendes Prinzip in der Biothik. In: Knoepffler/Haniel (ed.) Menschenwürde und medizin-ethische Konfliktfälle, 2000, p. 227.

Simon, J. and Taeger, J. Rasterfahndung. Eine kriminalpolizeiliche Methode. Baden-Baden, 1981.

Sokol, B. Gesundheitsdatenbanken und Betroffenenrechte: Das isländische Beispiel. In: NJW 2002, 1767 (1768).

Vitzthum, W. Graf. Rechtspolitik als Verfassungsvollzug? Zum Verhältnis von Verfassungsauslegung und Gesetzgebung am Beispiel der Humangenetik-Diskussion. In: Keller/Günther/Kaiser (eds.), Fortpflanzungsmedizin und Humangenetik, Tübingen, 1991.

Wiese, G. Gibt es ein Recht auf Nichtwissen? In: Jayme (ed.) Festschrift für Hubert Niederländer, Heidelberg 1991, p. 475.

Wiese, G. Genetische Analysen bei Areitnehmern. In: DÄBL. 1992, 656.

Wertz, D., Nippert, I., Wolff, G. and Aymé, S. Ethik und Genetik aus der Patientenperspektive: Ergebnisse einer internationalen Studie. Genomexpress 2/01.

PART IV
GENETIC SCREENING

10 Genetic and Nongenetic Medical Information: Is there a Moral Difference in the Context of Insurance?

Veikko Launis

Modern human genetics and its application has often been criticised by saying that it provides a new basis for discriminating against individuals, comparable in some important respects to the more established forms of human discrimination, notably racism and sexism.[1] These critics are referring to the continuing discoveries concerning people's genetic makeup. While the ground in itself is not totally new (as we know, race and sex are also genetically determined characteristics), many of the ethic issues and concerns it raises certainly are. Among the most serious concerns expressed by the critics is the fear that, if third parties such as insurance companies and employers are given an access to individuals' genetic information, this will inevitably lead to widespread social and economic discrimination and stigmatization.

The underlying idea of such criticism seems to be that genetic (test) information is somehow exceptional and should therefore be treated more carefully than other type of medical information. To the question what, if anything, makes genetic information morally distinguishable from all other health-related information, one of the following two answers is usually given. Adapting current terminology, I shall refer to these answers as *strong* and *weak genetic exceptionalism*. The strong form of genetic exceptionalism claims, roughtly, that genetic information is exceptional *per se*, that is inherently or qualitatively 'sufficiently' different from other kinds of health-related information that it deserves special protection or other exceptional measures.[2] The weak version denies this, but accepts that there may be something morally special about *the use* of genetic information for certain practical purposes, such as insurance and employment discrimination, which justifies exceptional treatment.[3]

It is essential to understand whether genetic information is indeed exceptional in order to evaluate current legislative and policy initiatives that distinguish genetic information from other medical information. In this chapter, I shall begin with an exploration of the concept of strong genetic exceptionalism. I shall consider three arguments about why genetic (test) information is inherently distinguishable from all other medical information and shall argue that all of these arguments fail. I shall then explore the weaker (and more restricted) claim that genetic test information is sufficiently exceptional with regard to its use for insurance purposes that it deserves special protection.

Strong genetic exceptionalism

The exceptional status that may be accorded to human genetic information can be seen to arise in numerous ways. Consequently, there has been considerable debate about the nature of such information, and different lists of reasons supporting genetic exceptionalism have been offered. To take some examples, according to George Annas, Leonard Glantz and Patricia Roche, genetic information can be considered fundamentally unique for at least three reasons: it can predict an individual's likely medical future for a variety of conditions; it divulges personal information about one's parents, siblings, and children; and it has historically been used to stigmatize and victimize individuals.[4] For David Orentlicher, again, genetic information is distinctive because [g]enetic makeup is at the hearth of personality. Genetics not only has a profound influence on such physical characteristics as height, weight, skin colour, and eye colour, but it almost certainly affects less tangible traits, such as shyness, altruism, sociableness, artistry, and intellectual skills.[5] Lori Andrews maintains, in the same spirit, that because genes are usually viewed as immutable and central to the determination of who a person is, information about genetic mutations may cause a person to change his or her self-image and may alter the way others treat that person.[6]

Are such distinctions persuasive enough to justify strong exceptionalism? I think we can offer a proper answer to this question only by providing a careful reconstruction of the argument. I suggest the following reconstruction.[7]

1) *Genetic information is predictive* Genetic tests provide more accurate data on the likelihood of an individual developing a particular medical condition than other medical tests.
2) *Genetic information is other-regarding* Genetic tests can provide (either certain or probablistic) information about the family members of the individual who is tested, and conversely, the health status of family members can provide information about the genetic constitution and future health of the individual.
3) *Genetic information is essential* Genetic information is more profoundly personal than other medical information, and one(s) essential identity is largely determined by one(s) genetic makeup.

Let me examine each of these distinctions in turn.

The predictive accuracy of genetic test information

On closer examination, the claim that genetic test information is more predictive than other medical information becomes less convincing. First of all, genetic tests can provide more accurate information about the subject's future health status only in the case of certain relatively simple single gene disorders, such as Huntington's disease and cystic fibrosis, and even in these cases the genetic tests are not considered to be completely foolproof. In more complex and, from the viewpoint of insurance companies, more interesting disorders such as Alzheimer's disease and certain types of cancer, the predictive accuracy of the genetic tests is considerably less, estimated to be of the same or even lower accuracy of prediction as standard medical tests (such as the test for LDL-cholesterol level which is often very predictive of the person's risk of coronary heart disease).[8] It seems, then, that the widespread belief that the predictive power of genetic tests will be high simply because DNA technology is involved is mistaken.

Secondly, and more importantly, it should be recognised that the (alleged) higher accuracy of prediction of genetic test information can hardly make it *morally* distinctive (and *ipso facto* morally more protectionable). On the contrary, it is plausible to hold that, all else being equal, the more accurate (predictive, reliable) the information, the more morally justified it is to incorporate it into actuarial calculations. That is to say, insurers should discriminate as accurately as they can.[9] This *prima facie* moral obligation stems from the principle according to which it is (other things being equal) more unfair to discriminate between individuals on statistical grounds than on individual grounds.[10] As Peter Singer puts it:

> To be judged merely as a member of a group when it is one's individual qualities on which the verdict should be given is to be treated as less than the unique individual that we see ourselves as.[11]

The real problem, then, is not that the distinction under consideration is scientifically too vague, but that, if solid, it will *weaken* rather than strengthen the moral basis of strong genetic exceptionalism.

Genetic test information and concern about others

As regards the second distinction (concerning the effects of genetic knowledge of one's family members), it has been pointed out that the history of familial disease has long, if not always, been a major component of the medical history of an individual, and the insurance industry has long consulted family medical histories when calculating premiums and determining insurance coverage. Access to family medical histories allows insurance companies to obtain actuarially relevant genetic information about prospective clients without the use of genetic testing. Thus, while some genetic information may be determined only through genetic tests, such tests are not the only, or even the principal, means of utilising the medical history of family members for the prediction of the disease status of an individual.[12]

For different reasons, I agree that the present distinction is fallacious. What is of importance is not that genetic differences are already taken into account in insurance decision-making (by taking into account family medical histories), but that genetic information falls under the broader category of *other-regarding* (medical) information and should be treated on a par with any other information belonging to that category. It should be clear from the outset that the mere fact that one person happens to be a *family member* of another carries no moral weight here. What, then, is left of the distinction? I think that what remains is the fact the people are sometimes connected with each other in such a way that medical information about one person can provide relevant medical information about another, and vice versa. But this is surely not a characteristic exclusively to genetic information. Consider, for example, someone who has a permanent relationship and who receives a positive result from a HIV test. Presumably, that piece of medical information can (and under certain circumstances will) provide knowledge about the person's sexual partner. Likewise, as Thomas Murray points out,

> [t]hat one member of a family has tuberculosis is certainly relevant to the rest of the household, all of whom are in danger of infection, along with everyone who works with or goes to school with the infected individual.[13]

It seems then, that the distinction between genetic and non-genetic medical information draws the line in the wrong place. Genetic information may be more frequently other-regarding than non-genetic medical information, but surely not always, and not in a sense that would make it, or contribute to its being, exceptional in the strong sense.

Genetic essentialism

The third proposed distinction is presupposed by the idea that the genetic contribution to a multifactorial medical condition is more fundamental than, and can be readily separated from, the non-genetic contribution, and hinges on the deeply rooted cultural belief that genetic information represents an individual's immutable and fundamental characteristics and is therefore more essential to their nature than other types of medical information. As Joseph S. Alper and John Beckwith have pointed out, there is a strong tendency among the general public to regard people's genotype as the ultimate determinant of their nature. Since we have no choice in selecting our genotype, nor control over the expression of it, discrimination on the grounds of genetic information is felt to be more unfair than discrimination on grounds of non-genetic information. According to Alper and Beckwith, such a way of thinking is mistaken because it oversimplifies and distorts the very complex relationship between human nature, human genome, and the environment.[14]

This way of thinking is especially challenging from the bioethical viewpoint and deserves to be considered in more detail. According to this view, genetic information can be distinguished from non-genetic medical information with respect to its closer relationship to people's essential identity or nature. There seems to be more than just a cultural belief at issue here. The main reason for claiming that genetic characteristics are more fundamental to our essential identity or nature than non-genetic characteristics can be expressed in the form of the following argument:

P1: C (some human characteristic) is constitutive of the essential identity or nature of X (some human being) if and only if C is immutable.
P2: All genetic characteristics of X are immutable.
Hence,
P3: All genetic characteristics of X are constitutive of the essential identity or nature of X.

Note that this essentialist argument may be sound even if the distinction between genetic and non-genetic medical information that hinges on it turned out to be fallacious, because it is not impossible that there are some non-genetic medical characteristics that are immutable as well.[15] However, the converse does not hold: if the argument fails, then there is little reason to adhere to the distinction.

For the sake of argument, let us suppose there is such a thing as 'essential identity' or 'core human nature'.[16] A crucial question, then, is this: What sorts of characteristics identify a person as essentially the person he or she is, such that if those characteristics were changed, he or she would be a significantly different person?[17] This question expresses what may be called the formal definition of essence (or essential characteristics). It says that our individual essence is constituted by characteristics which are such that they cannot be altered or removed without, metaphysically speaking, our ceasing to exist. Now, according to P1, what is materially speaking distinctive in such characteristics is that they are *immutable*.[18] What precisely does this mean? In the light of what has been suggested in the literature, the expression 'C is immutable' might be taken to mean any of the following:

1) C cannot be altered.
2) C is not susceptible to alteration.
3) C cannot be altered by X.

Of these, (1) expresses strong immutability, because it allows no change in C. Options (2) and (3) express weaker immutability, the former by granting that C may occasionally or, to some extent (for example, in the course of natural evolution), be capable of change, and the latter by granting that some agent other than X (e.g. a group of eminent geneticists) may be empowered to change C.

What should we think of these options? It seems to me that they are *all* counterintuitive in the present context. Consider first (1). It implies that if C constitutes part of X's essence at t_1 and becomes alterable at t_2, then it can no more be constitutive of X's essence. This is implausible. Think for instance of the possibility of undergoing a full sex change by cosmetic surgery and hormonal treatment. Despite this, the majority of us continue to believe that gender is a constituent of our essential identity. Should we abandon this belief *just because* it has now become possible to exercise control over gender? I do not think so.

Consider next (2). It is at least as suspect as (1), because it implies that if C (say the property of becoming bald at some point of life) constitutes part of X's essence at t_1 and becomes more controllable at t_2, then it can no more be constitutive of X's essence. Finally, (3) is also suspect, because it too implies that if X and (some other person) Y both have the same characteristics C, and X has the power to change it while Y has not, then C can be constitutive only of Y's essence.

It appears, then, that if my above analysis is correct, that 'C is immutable' amounts to nothing more than its formal definition, namely

4) C cannot be altered without X's ceasing to exist.

According to (4), immutable characteristics are simply characteristics which people cannot lack without losing themselves. I believe this is the only consistent way of understanding the 'immutability' of essential identity. My analysis does not, of course, show that P1 is false. What it shows is that the premise is problematic in the sense that it is 'empty' and provides thus no *material* criterion for distinguishing between essential and nonessential medical characteristics.

Let me turn now to P2. It claims that our genetic characteristics are immutable, which should now be taken to mean nothing more than that they cannot be altered without our ceasing to exist. A solid metaphysical principle that would seem to support this interpretation has been offered by Saul Kripke. In discussing the logical conditions of identity Kripke argues as follows:

> How could a person originating from different parents, from a totally different sperm and egg, be *this very woman* [Elizabeth II]? One can imagine, *given* the woman, that various things in her life could have changed: that she should have become a pauper, that her royal blood should have been unknown, and so on. ... This seems to be possible. And so it's possible that even though she were born of these parents she never became queen. ... But what is harder to imagine is her being born of different parents. It seems to me that anything coming from a different origin would not be this object.[19]

In other words, genetic characteristics are immutable in the sense that we cannot have had a *genetic origin* other than our actual genetic origin. Kripke seems to hold that the (logical) relationship between a particular genotype and a particular essential identity is asymmetrical: while no-one could have developed from a genotype other than the one from which he or she did develop, it does not follow that he or she is the only person who could have developed from that genotype.

Now the essential question is: how far does this bring us? I believe not far enough to establish the truth of P2. Notice, first, that Kripke's 'necessity of origin' principle does not entail that *any* subsequent change in one's initial genetic constitution would destroy one's essence.[20] As Robert Elliot has explained:

> Consider a particular table. It is the particular table that it is because it has a particular origin: its initial constitution fixes it as *this* particular table. Subsequently one of the table's legs is replaced. Does this replacement disrupt the identity of the table? No, it is still the same table: it survives without continued possession of all original parts. Likewise with organisms ... The initial instantiation of a particular genotype makes [an organism] the organism that it is. But this is consistent with the claim that subsequent alterations to that genotype, or the material which embodies it, do not necessarily destroy [the organism].[21]

Presumably, whether or not the replacement of genetic material destroys one's essential identity depends at least in part on *how much* and *what kind* of genetic material is replaced. The substitution of most (say 80-90%) of the functional genetic material[22] (including the material that is responsible for one's mental characteristics) would probably constitute one's cessation.[23] But what about smaller, more realistic, proportions?

It is implausible that the substitution of say one or two normal genes for abnormal ones (as could happen in human gene therapy) would necessarily destroy one's essential identity. First of all, not all genetically determined characteristics are essential. For example eye-colour, which is genetically determined, is not usually regarded as an essential characteristic of a human being.[24] This is also true of many genetically determined disease characteristics.

Secondly, while there are many essence-constituting characteristics which have a genetic basis, it does not follow that to learn the genetic basis of such characteristics would be to know one's (genetic) identity or essence. This is true for the following reason: let G be some set of genetic characteristics (e.g. certain high risk variants of the insulin gene), C some set of genetically determined essence-constitutive characteristics (e.g. type 1 diabetes mellitus) and X the bearer of these characteristics. Suppose, then, that some alteration in G (say the substitution of a normal insulin gene for the defective one) would necessarily change X's essential identity. Does it follow that, necessarily (i.e. irrespectively of what X thinks), G is X's essential set of characteristics? No, it does not. This is because nothing can be essential only in virtue of being causally related to essential characteristics; causal connection is too weak a relationship for transferring essence. In order for such a transference to occur, the connection between G and X's essence-constituting characteristics C would have to be conceptual, not merely causal. In most (if not all) actual cases, however, the connection is clearly causal.[25]

The objection could be raised that a causal relationship is enough to make G X's essential characteristics, if X is psychologically unable to conceive of C without thinking either of G, or the fact that G is causally responsible for C, as *part* of C. According to this objection, X's position may be seen as analogous to the position of an agent who aims at A and must also aim at B (as part of A), because he or she knows that B is a prerequisite of A.[26]

This claim can be countered as follows. Firstly, that could be the case only if C was causally determined by G (i.e. if G was a causally necessary and sufficient condition of C), for otherwise X's regarding G as essential could be best explained by X's thinking of G as a set of *genetic* characteristics rather than as one which causally contributes to C. If both genetic and non-genetic factors are causally responsible for C, why should only genetic factors be considered essential? Perhaps because they are *explanatory* causal factors. As Carl F. Cranor points out, explanatory causal factors are factors which are related to contingencies that do not ordinarily occur and can therefore account for the departure from the normal course of events. But *why* would only genetic causal factors be explanatory? I think that no satisfactory answer can be given to that question. I think that Cranor is perfectly right in maintaining that '[w]hich contingencies we select from the set sufficient to produce the event depend upon the context and our interests'.[27] It is hard to provide much in the way of objective guidance for selecting 'the or a cause of a particular outcome'.

Secondly, and more importantly, the objection would make sense only if a distinction was made between perfect and imperfect (or derived) essential characteristics, the former being characteristics which are or can be essential *per se*, the latter being those characteristics which are or can be essential only through necessitating certain perfect essential characteristics. However, drawing such a distinction makes no sense, because the notion of imperfect essential characteristics is completely unintelligible.

A third reason for thinking it implausible that the substitution of a limited number of normal genes for abnormal ones would necessarily destroy one's essential identity is related to the implausibility of objective essentialism: it seems incontestable that which genetic or genetically determined characteristics are essential to our being will depend, at least to some extent, on us. As David Heyd has remarked, what makes people essentially what they are depends on what, from their point of view, 'would be considered a change so radical in their life or character that *they* would not consider the result to be themselves any more'.[28] In the final analysis, we are the only ones who can decide whether the consideration that we have some genetic or genetically determined characteristics has some special significance for us.

The above considerations do not, of course, exclude the possibility that some genetic or genetically determined characteristics turn out to be (subjectively) essential to some people. (Presumably, as Thomas Murray points out, the more we repeat that genetic information is fundamentally unlike other kinds of medical information, the more support we implicitly provide for genetic essentialism.)[29] Nor do they exclude the possibility that – perhaps due to some characteristics of genetic information or of the society into which it will flow – the idea of genetic (or genetically determined) characteristics as essential characteristics becomes a powerful social construction with widely harmful social effects.[30] The upshot of my argument is merely that genetic characteristics need not have any more importance for us than any other characteristics. This is however, enough to show that P2 is at least suspicious.

To conclude, then, strong genetic exceptionalism is hard to

defend. Given the many difficulties in distinguishing genetic (test) information from non-genetic medical information, it would be inconsistent to prohibit insurance companies from using genetic test results in fixing premiums and yet to allow the use of non-genetic test results for the same purpose *without a further moral reason*. The widely held cultural belief or metaphysical conviction, according to which genetic information is more intimately related to people's essence than any other medical information, if set forth as an objective statement, is simply incorrect and may perhaps as such be better understood as a distorted reflection of the metaphysical ideas (inherent in our culture) captured by Kripke's necessity of (genetic) origin principle.

Weak genetic exceptionalism and insurance discrimination

Weak genetic exceptionalism claims that there may be something morally special about the use of genetic test information for certain practical purposes, especially for insurance underwriting. Contrary to this claim, some spokespersons of insurance companies have argued that there is nothing exceptional about the use of genetic information for such purposes. Differentiation based on genetically-determined characteristics has been an approved practice in insurance business for quite a long time. Thus, the familial inheritance of may well-known genetic disorders such as Huntington's disease and cystic fibrosis is already taken into account in determining whether, or on what terms, to offer coverage to individuals. If an individual's genetic make-up will affect the quality or length of his or her life, then an insurer has an unquestionably legitimate interest in knowing so. The consideration that the information can now be obtained by genetic testing does not in any way alter this fact.[31] It may even be claimed that discrimination among individuals is (and should be) *intrinsic* to the nature of insurance. The real business of insurance is dependent upon accurate risk classification and differentiation, founded on the idea that the premiums should be more or less in proportion to the estimated risk level of policy holders. According to this view, any questioning of that principle would undermine the basic idea of private insurance.[32]

This is a telling argument. However, it should be observed that there are other, more compelling moral reasons for thinking that those of us who are genetically disadvantaged should have access to insurance services on roughly the same terms as the genetically 'normal' or advantaged.[33] The following paragraph by Gregory S. Kavka captures some of them:

> [S]ociety and its members benefit from the feeling that we are all together in a common enterprise and share one another's fate. Insurance can be a device for forcing us to share one another's fate: through it, uncertainty, self-interest, and risk aversion combine to make us take care of one another to some extent. If too much uncertainty is replaced by knowledge – as in the case of genetic information about illness susceptibility – this useful device may be threatened ..., and will need to be repaired and rescued.[34]

Kavka's point is that uncertainty about genetic risks is socially beneficial; it promotes social solidarity by 'preventing the most severely affected victims of any social policy from being preidentified'.[35] Uncertainty about genetic risks allows the members of the group or community to share the sense that 'we are all in this together', and a belief that the policies are chosen to promote general welfare. This would not be possible if policies were founded upon preidentified certain 'genetic victims' by preidentified certain 'genetic beneficiaries'. In so far as our knowledge of individual genetic risks becomes more certain (and less private), the system of solidarity represented by the insurance industry will be undermined.[36]

Thus, instead of minimization of risk, insurers should focus their activities on assessment of risk, since this is their primary (solidarity-sustaining) social function, at least to the extent that this is economically viable.[37] If they do not do this, the insured group will consist primarily of low risk policy holders, and very soon we will have moved from 'insurance against all risks to insurance only of the risk free'.

By way of concluding, then, I have argued that, given the many difficulties in distinguishing morally between genetic and non-genetic medical data, prohibiting insurance companies from using genetic test results in fixing premiums whilst permitting the use of non-genetic (medical) test results for the same purpose *without a further moral reason* would be inconsistent. However, I have tried to show that there are good solidarity-based reasons for regarding insurance discrimination on the basis of genetic test information as morally unacceptable. Our sustaining genetic uncertainty on the collective level has morally valuable consequences, which are not easily overridden by other types of moral considerations.[38]

Notes

1. Wolf, Susan, M. (1995) 'Beyond AGenetic Discrimination: Toward the Broader Harm of Geneticism', *Journal of Law, Medicine and Ethics*, 23, p.345.
2. Murray, Thomas H. (1997) 'Genetic Exceptionalism and AFuture Diaries: Is Genetic Information different from other medical information?' in Mark A. Rothstein (ed.) *Genetic Secrets: Protecting Privacy and Confidentiality in the Genetic Era*. Yale University Press, New Haven and London.
3. See ibid. p.64. Murray's distinction is not precisely the same as mine but comes very close to it.
4. Annas, G.J., Glantz, Leonard H. and Roche, Patricia A. (1995) 'Drafting the Genetic Privacy Act: Science, Policy and Practical Considerations', *Journal of Law, Medicine and Ethics*, 23, p.360.
5. Orentlicher, D. (1997) 'Genetic Privacy in the Patient-Physican Relationship' in Mark A. Rothstein (ed.) *Genetic Secrets: Protecting Privacy and Confidentiality in the Genetic Era*, Yale University Press, New Haven and London, pp.79-80.
6. Andrews, Lori B. (1997), 'Gen-Etiquette: Genetic Information, Family elationships and Adoption' in Mark A. Rothstein (ed) *Genetic Secrets: Protecting Privacy and Confidentiality in the Genetic Era*, Yale University Press, New Haven and London, p.255. For further suggestions, see Joseph S. Alper and John Beckwith, 'Distinguishing Genetic from Nongenetic Medical Tests: Some Implications for Antidiscrimination Legislation', *Science and Engineering Ethics*, 4 (1998), pp.141-150; Søren Holm, 'There is Nothing Special about Genetic Information' in Alison K. Thompson and Ruth F. Chadwick (eds) *Genetic Information: Acquisition, Access and Control* (New York: Fluwer Academic/Plenum Publishers, 1999), pp.97-103; Human Genetics Commission, *Whose Hands on Your Genes? A discussion document on the storage, protection and use of personal genetic information* (London: Human Genetics Commission, November 2000), pp.6-7; Martin Richards, 'How Distinctive is Genetic Information?' *Studies in History and Philosophy of Biological and Biomedical Sciences* 32C (2001), pp.663-687; Lainie

Friedman Ross, 'Genetic Exceptionalism vs. Paradigm Shift: Lessons from HIV', *Journal of Law, Medicine and Ethics*, 29 (2001), pp.141-148.

7. Cf. Alper and Beckwith, 'Distinguishing Genetic from Nongenetic Medical tests: Some implications for antidiscrimination legislation', 143.
8. Ibid., 144; Abby Lippman, 'Led (Astray) by Genetic Maps: The Cartography of the Human Genome and Health Care', *Social Science and Medicine*, 35 (1992), pp.1470-1471; Holm, 'There is nothing special in genetic information', 99.
9. Among the few who deviate from this principle is R.L. Zimmern who writes, 'The morally relevant circumstances in which regulation is warranted [include] were the predictive value of a test and the probability of developing the disease, or phenotypic manifestations, attributable to the genetic defect is high enough to give ... insurers or others in society a reason to justify disciminatory policies. R.L. Zimmern 'Genetic Testing: A conceptual exploration', *Journal of Medical Ethics*, 25 (1999), p.153.
10. Discrimination is called statistical when some human characteristic X is positively, but imperfectly, correlated with some other human characteristic Y and one discriminates between individuals on the basis of X, X being used as a proxy for Y. Correspondingly, discrimination can be called individualised when one discriminates between individuals either directly on the basis of Y, or on the basis of X when there is a perfect positive correlation between X and Y. See Stephen Maitzen, 'The Ethics of Statistical Discrimination', *Social Theory and Practice*, 17 (1991), p.23.
11. Singer, Peter 'Is racial discrimination arbitrary?', *Philosophia* 8 (1978), p.195. See also Joel Feinberg, *Harm to others* (New York and Oxford: Oxford University Press, 1984), pp.199-202; Perry C. Beider, 'Sex discrimination in insurance', *Journal of Applied Philosophy* 4 (1987), pp.66-68; Paul Fenn and Stephen Diacon, 'Disability and Insurance' in Tom Sorell (ed.) *Health Care, Ethics and Insurance* (London and New York: Routledge, 1998), p.128.
12. Alper and Beckwith, 'Distinguishing Genetic from Nongenetic Medical Tests: Some Implications for Antidiscrimination Legislation' p.144; Zimmern, 'Genetic Testing: A Conceptual Exploration', p.152.
13. Murray, 'Genetic Exceptionalism and AFuture Diaries: Is Genetic Information Different from Other Medical Information?', p.65. Additional examples are provided by Søren Holm (1999, p.100) who writes: 'The birth of a child with the stigma of congenital syphilis is highly informative about its parents, and the finding that one member of a family suffers from low-grade lead poisoning is indicative of the same condition in other family members'.
14. Alper and Beckwith, 'Distinguishing Genetic from Nongenetic Medical Tests: Some Implications for Antidiscrimination Legislation' pp. 143-144; cf. Max Charlesworth, 'Human Genome Analysis and the Concept of Human Nature' in Derek Chadwick, Greg Bock and Julie Whelan (eds) *Human Genetic Information: Science, Law and Ethics* (Chichester: John Wiley and Sons, 1999), p.192; Lippman, 'Led (Astray) by Genetic Maps: The Cartography of the Human Genome and Health Care', p.1470; Jacqueyn Ann K. Kegley,

'Genetic Information and Genetic Essentialism: Will We Betray Science, the Individual and the Community?' in Jacquelyn Ann K. Kegley (ed.) *Genetic Knowledge: Human Values and Responsibility* (Lexington: An ICUS Book, 1998), p.49-50; Angus Clarke, 'The Genetic Dissection of Complex Traits' in Veikko Launis, Juhani Pietarinen and Juha Räikkä (eds) *Genes and Morality: New Essays* (Amsterdam and Atlanta: Rodopi, 1999), pp.114-116.
One must not confuse 'genetic essentialism' with 'genetic determinism', as is sometimes done. Genetic essentialism states, roughly, that genetic characteristics are more essential to our nature than other (non-genetic) characteristics, whereas genetic determinism is a doctrine according to which human behaviour is rigidly determined by genes and cannot be changed by environmental factors. There is no logical connection between the two doctrines; nor is there any logical connection between genetic non-essentialism and 'radical environmentalism', i.e., the view that the only factor that influences an individual's phenotype is its environment. For a detailed account of genetic determinism, see e.g. Sahotra Sarkar, *Genetics and Reductionism* (Cambridge: Cambridge University Press, 1998), pp.10-13.

15. See e.g. Murray, 'Genetic Exceptionalism and AFuture Diaries: Is Genetic Information Different from Other Medical Information', pp.65-66.
16. Derek Parfit, for one, has argued that there is no such thing. See his *Reasons and Persons* (Oxford: Clarendon Press, 1984), pp.204-209, 223.
17. Cf. Amelie Oksenberg Rorty, 'Introduction', in Amelie Oksenberg Rorty (ed.) *The Identities of Persons* (Berkeley and Los Angeles: University of California Press, 1976) pp.1-2.
18. In their *From Chance to Choice: Genetics and Justice* (Cambridge: Cambridge University Press, 2000), p.87, Allen Buchanan, Dan W. Brock, Norman Daniels and Daniel Wikler write: 'Human nature has traditionally been regarded ... as unchanging. So the possibility of changing even our Aessential characteristics would seem to render the very concept of human nature obsolete, so far as it includes the idea of an unchanging core of essential characteristics'.
19. Kripke, Saul A (1980) *Naming and Necessity*, Harvard University Press, Cambridge, Mass, p.113.
20. Stated more succinctly, the principle runs as follows: '*If a material object has its origin from a certain hunk of matter, it could not have had its origin in any other matter*'. Ibid., 114fn.
21. Elliot, Robert (1993) 'Identity and the Ethics of Gene Therapy', *Bioethics*, 7, pp.30-31.
22. There are large parts of the human genome, known as 'junk DNA', that appear without a clear functional purpose. A new estimate shows that the number of human genes is considerably smaller than was previously estimated. A human genome is now estimated to contain between 28,000 and 34,000 genes, while previous estimates ranged from 60,000 to 140,000. See Guido Pincheira, 'The Human Genome: Facts, Enigmas and Complexities' in Kegley (ed.) *Genetic Knowledge: Human Values and Responsibility*, p.11.

23. Elliot, 'Identity and the Ethics of Gene Therapy', 31; Noam J. Zohar, 'Prospects for A Genetic Therapy – Can a person benefit from being altered?' *Bioethics*, 5 (1991), pp.285-287.
24. Zohar, N.J. (1991) 'Prospects for AGenetic Therapy – Can a person benefit from being altered?' *Bioethics*, 5, pp.285-286.
25. Cf. Buchanan et al., *From Chance to Choice: Genetics and Justice*, 85, p.160.
26. I am grateful to Olli Koistinen for pointing this objection out to me.
27. Cranor, Carl F. (1994) 'Genetic Causation' in Carl F. Cranor (ed.) *Are Genes Us? The Social Consequences of the New Genetics*, Rutgers University Press, New Brunswick, p.128.
28. Heyd, David (1992), *Genethics: Moral Issues in the Creation of People*, University of California Press, Berkeley and Los Angeles, pp.165-166; cf. Charlesworth, 'Human Genome Analysis and the Concept of Human Nature', p.187.
29. Murray, 'Genetic Exceptionalism and AFuture Diaries: Is Genetic Information Different from Other Medical Information?', p.71.
30. Ibid. pp.60-73; Draper, Elaine (1991), *Risky Business: Genetic Testing and Exclusionary Practices in the Hazardous Workplace*, Cambridge University Press, Cambridge; Richards, 'How Distinctive is Genetic Information?', pp.671-674.
31. Jaeger, Ami S. (1993), 'An Insurance View on Genetic Testing', *Forum for Applied Research and Public Policy*, p.25.
32. See Murray, Thomas H. (1992), 'Genetics and the Moral Mission of Health Insurance', *Hastings Center Report*, 22, pp.14-15; Harper, Peter S. (1993) 'Insurance and Genetic Testing', *The Lancet*, 341, p.224.
33. For the full argument, see my 'Solidarity, Genetic Discrimination and Insurance: A Defence of Weak Genetic Exceptionalism', unpublished manuscript.
34. Kavka, Gregory S. (1994) 'Upside Risks: Social Consequences of Beneficial Biotechnology' in Carl F. Cranor (ed.) *Are Genes Us? The Social Consequences of the New Genetics*, Rutgers University Press, New Brunswick, p.176.
35. Kavka, Gregory S. (1990) 'Asome Social Benefits of Uncertainty', *Midwest Studies in Philosophy*, 15, p.312.
36. Ibid.
37. I have argued elsewhere that adverse selection (the tendency of those who present a poorer-than-average risk to apply for insurance to a greater extent than do people with average or better-than-average expectations of loss) in the genetic area poses no special problem for insurance companies, as long as the possibility of potential adverse selectors to purchase unusually large (life) policies is limited. See my *Multidimensional Bioethics: A Pluralistic Approach to the Ethics of Modern Human Biotechnology* (Turku: Reports from the Department of Philosophy of the University of Turku, vol. 6, 2001), pp.51-56.

38. This article has been developed out of previous material written by the author. I am grateful to Opragen Publications for permission to reproduce extracts from 'The Use of Genetic Test Information in Insurance: The Argument from Indistinguishability Reconsidered', *Science and Engineering Ethics* 6 (2000): pp.299-310.

11 New Practices of Screening in the Field of Cancergenetics: A Co-evolutionary Perspective

Dirk Stemerding and Annemiek Nelis

Introduction[1]

New genetics is used as a catchword referring to the development and introduction of a variety of new genetic technologies and practices in many different fields of society. As a particular significant characteristic of the emerging new genetics, we see the extension of options for predictive genetic testing from classic and relatively rare hereditary disorders to common diseases like cancer (Bourret *et al.* 1997). This development has aroused new hopes and expectations about more effective forms of healthcare as well as growing concern about its wider ethical and social implications. Talking about the new genetics as 'arousing' new hopes, expectations and concerns, easily suggests a linear cause-effect relationship in which hopes, expectations and concerns are only responses from society to new technological developments. This however is only (a too small) part of the story. The emergence of new genetics in society can more adequately be described as a process of 'mutual shaping' in which new technical options, practices, hopes, expectations and concerns are developing together in an interactive way. In science and technology studies such processes of mutual shaping are often understood in terms of 'co-evolution' (Rip and Kemp, 1998).

In this chapter we focus on cancergenetics as a field in which the extension of options for genetic screening is clearly visible and which allows us to study the emergence and implications of the new genetics from a co-evolutionary perspective. From this perspective two points are especially important for our understanding of the new genetics in society.

First, co-evolution is a process in which technology and society are mutually shaped, starting with new options, initiatives and expectations ('varieties') on a local level. Such local activities may then evolve and stabilize on a more collective level in new configurations of artefacts, practices, professions, users, rules, institutions, etc. In this way local screening initiatives in the field of cancergenetics have taken shape in more widespread practices of detection of hereditary cancers, involving the establishments of new risk groups, new institutions, new choices and new responsibilities. As a result, patterns will occur in processes of co-evolution that shape – enable and constrain – further actions, interactions and developments.

Second, the development and embedding of new technologies in society will always be conditioned – enabled and constrained – by alignments and arrangements resulting (as 'selection environment') from earlier processes of co-evolution of technology and society. The new genetics for example is developing in a 'socio-technical' landscape of existing clinical practices, in which clinical genetics has obtained a special place, offering a particular setting of non-directive counselling for various forms of genetic screening (Nelis, 1999). Thus, new screening initiatives in the field of cancergenetics will have to meet, or indeed challenge, conditions embodied in the alignments and arrangements of established clinical practices like clinical genetics.

These two points we will consider in more detail in the following account of how in the Netherlands a screening practice has emerged for Familiar Adenomatous Polyposis (FAP), a particular hereditary predisposition for colon cancer. First we will see how in the 1980's screening of individuals at risk became a regular practice on a national scale. Then we describe how this screening practice, with the development and introduction of DNA-diagnostic tests, began to interfere with the established practice of clinical genetics. In conclusion we will discuss the implications of our case-study in the wider context of the emerging new genetics in society.

FAP in the early eighties: the emergence of a screening practice

In 1981 a collection of three articles and a commentary appeared in one of the issues of the Dutch Journal of Medicine, focusing on the clinical experiences with a rare, dominantly inherited disease, called polyposis coli or familiar adenomatous polyposis (FAP). In each of the three articles an elaborate description was given of the history of the disease in a particular family. One of the articles opens with the story of a 27-year old woman who consults the clinic because a 34-year old cousin of hers had been recently identified as a FAP patient after diagnosis of colon cancer (De Ruiter and Den Hartog Jager, 1981). The mother of the woman had died from colon cancer when she was 43, a few years after the colon partly had been removed because of a malignant polyposis. With regret, the authors of the article observe that clinical examinations of relatives had not been undertaken at the time, although it was known that a grandmother and a great-grandfather also had died from "cancer of the colon". The woman who was consulting the clinic had no symptoms, but inspection of the colon revealed many polyps that resulted in the diagnosis of FAP and in the decision to completely remove the colon. The article then continues with the case of a younger sister of the woman, who likewise was without symptoms, but where after examination also polyposis was found and the colon removed. In other brothers and sisters examination only revealed a few polyps in the colon. In these cases, as the authors point out, the examination will have to be repeated every year. In conclusion, the article reports the results of examinations that had been carried out in 39 persons, spanning two generations of the family. In twelve cases polyposis was found. In four of them colon cancer had already developed. Four persons refused the invitation to undergo an examination.

The collective publication of the three articles and the accounts given by the authors of the diagnosis, treatment and screening in families where FAP is found, may be seen as an event which marks the emergence of a specific clinical practice. That is, a practice in which the professional responsibility of the medical specialist – an internist or gastro-enterologist – cannot be restricted to the individual patient, but should extend also to the health and survival of the patient's relatives (see Figure 11.1 below). When a patient consults the clinic with symptoms and the diagnosis FAP is made, in most cases a fatal colon cancer will have already appeared. Thus, as the authors of the articles point out, it is of vital importance to trace the families in which FAP is found and to screen the members of these families every two or three years, beginning from about age ten. As soon as in the colon more than one hundred polyps are found, the diagnosis FAP should be made, and it is only by complete removal of the colon that the development of cancer can be prevented. Even then, regular screening will remain necessary. When, on the other hand, members of the family are still free of symptoms between age forty and fifty, the appearance of polyposis can be reasonably excluded and screening may be terminated.

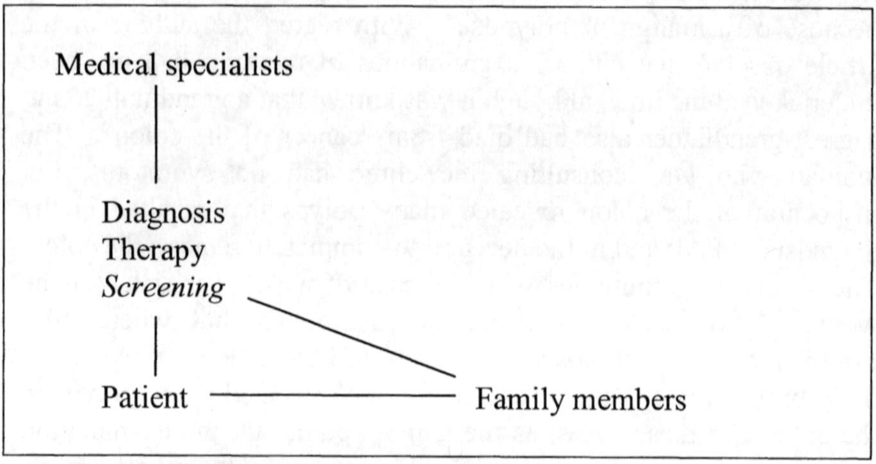

Figure 11.1 Emergent cancergenetic practice in the field of polyposis (late 1970s)

The conclusions and recommendations of the authors clearly were intended to promote a practice that, in the beginning of the eighties, was still in its infancy. At that time, it was already considered against due practice to refrain, when the diagnosis of FAP was made, from an extensive family anamnesis. However, in the period before, such a family anamnesis by no means was the rule, which in one of the articles is explained by a general lack of knowledge of the serious consequences of the disease. Even when an extensive mapping of the family history of the patient followed the diagnosis of FAP, it was often considered to be an impossible task for the individual specialist to actively approach all the family members involved. And, as far as family members were approached, the specialist could not always be sure that they would regularly return for a periodical screening. Thus, in a commentary on the three articles mentioned above, published in the same issue of the Dutch Journal of Medicine, it was observed that:

> Detective-like genealogical investigations, the psychological burden experienced by people who feel completely healthy and yet face the prospect of invasive examinations of the colon, and the not always interesting task to screen a fairly large number of people who have no symptoms, require a great and unremitting enthusiasm and dedication of those who undertake to follow a family with a history of polyposis.
>
> (Van Slooten, 1981)

Hence, the author of the commentary argued for the establishment of a centralized national registry, which could send out a reminder to medical specialists each time when a person at risk had to be called up for screening. The results of the screening would have to be returned to the registry. When no results followed, the organisation could take further action in order to safeguard the care for those at risk and to have certainty about their condition. With his plea, the author actually repeated a message that he had voiced already in the same journal no less than 25 years before. This time, however, the argument would find a hearing.

Late eighties: the establishment of a national registry

At the end of the eighties, a patient who consulted the clinic with symptoms of FAP would meet a practice that indeed was different from what we have seen before. Now, the medical specialist in attendance of the patient not only had to inform him or her about the hereditary nature of the disease and about the importance of screening family members, but also could refer the patient to the national *Foundation for the Detection of Hereditary Tumours*. In 1983, this foundation was established by a number of specialists involved in the treatment of patients and their families suffering from hereditary tumours (one of the founders was the author of the commentary mentioned above). In 1985 the Foundation started a national registry of families with a history of FAP. Thus the Foundation aimed to promote screening in high-risk families, to guarantee the continuity of screening, to collect data for scientific purposes, and to offer advice about diagnosis, treatment, methods of screening, and genetic services for counselling (Vasen *et al.* 1988).

The result of this development was a more extended practice of diagnosis, treatment and screening of FAP, in which every patient is reported by medical specialists to the Foundation for the Detection of Hereditary Tumours (see Figure 11.2 below). A social worker of the Foundation then approaches the patient and with his or her help draws up a family tree which makes it possible to trace the history of the disease and to identify members of the family who are at risk. The patient is asked to inform relatives at risk and to urge them to have themselves screened. If they agree, family members are approached by the Foundation with a request for registration. In this way, nearly all FAP families in the Netherlands have been registered, accounting for a few hundreds of families (Annual Report, 1994). Of those who have themselves registered, personal and medical information is collected and, through a system of recall, specialists are notified when individuals should be called up for screening. When no screening results are reported back and upon inquiry it appears that someone did not come up for screening, the registry will send out a request to the family doctor to take action and to remind this person that screening is of vital importance.

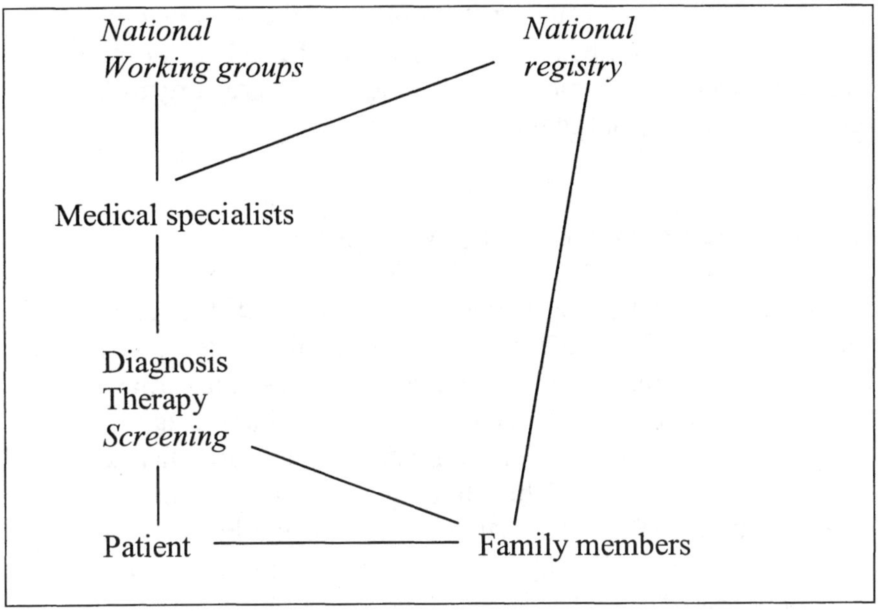

Figure 11.2 Establishment of a national registry in the field of polyposis (late 1980s)

While the care for individuals at risk of FAP at first strongly depended on the enthusiasm and efforts of individual specialists and on the consciousness of those at risk, it is now the Foundation for the Detection of Hereditary Tumours which has taken the responsibility for the organisation and continuity of screening and which 'will put all efforts in encouraging (registered) individuals to comply (to regular screening)' (Annual Report 1994, p. 5). Indeed, as those working for the Foundation point out, in order to motivate family members to participate in a screening programme, good information and a personal approach including visits at home are necessary (interview data, see note 6). Moreover, through the establishment of a national registry, information is collected which not only facilitates the organisation of screening programme, but which also creates possibilities for systematic follow-up and evaluation of its results. For that purpose, two national working groups on FAP, involving various forms of expertise, collaborate with the Foundation in the organisation of studies and the establishment of guidelines (Vasen *et al.* 1988). Thus, through the efforts of the Foundation, local practices of early detection and prevention have become part of a larger network in which these practices were organised and regulated on a national scale.

Early nineties: the advent of DNA-diagnosis

In the early nineties, a patient who consulted the clinic with symptoms of FAP again would meet a practice that was extended with new elements – new technologies, rules and organisations (Vasen and Müller, 1991). In 1991 molecular biologists succeeded to relate the occurrence of FAP to mutations in a particular gene, the so-called APC gene. This finding opened the possibility of presymptomatic DNA-diagnosis whereby on the basis of mutation analysis members of a FAP family may be informed about their individual risk-status, that is, whether they will get the disease or not. However, in the Netherlands, DNA-diagnosis is made available only through a network of regional clinical genetics centres. Thus, with the advent of DNA diagnosis of FAP, molecular biology laboratories and clinical genetic centres became part of the network in which the practice of diagnosis, treatment and screening of FAP took shape (see Figure 11.3 below).

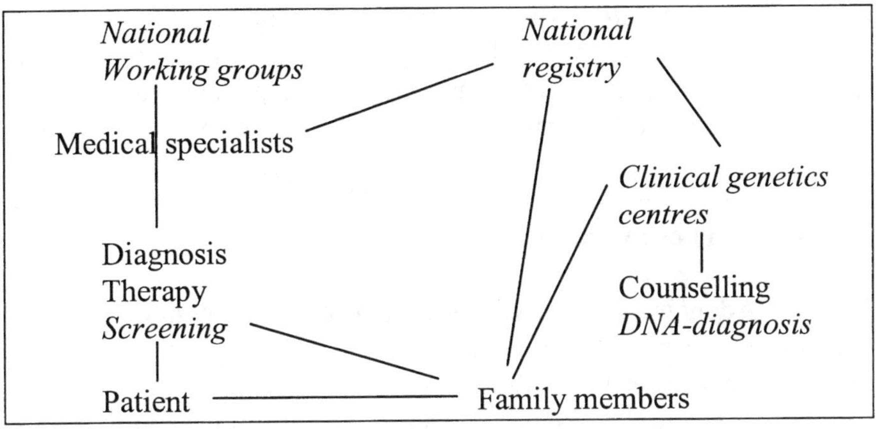

Figure 11.3 Introduction of DNA-diagnosis in the field of polyposis (1990s)

A new patient now not only will be reported by the medical specialist to the Foundation for the Detection of Hereditary Tumours, but also will be referred to a clinical genetics centre for mutation analysis. Again, a counselor of the clinical genetic centre will draw a family pedigree in order to identify relatives who may be at risk for the disease. Members of the family then have the opportunity (after having been informed by the patient) to be referred to a clinical genetics centre, which may offer them presymptomatic DNA-diagnosis as soon as a mutation is found. Those who accept the offer and are diagnosed as carriers, have the certainty that they will get FAP and that regular screening is the only way to escape from an early and deadly cancer. Of those diagnosed as carriers most, if not all will have themselves registered in the national registry of the Foundation for the Detection of Hereditary Tumours. Those however who are diagnosed as non-carriers are excluded from risk and thus may abstain from participation in a burdensome and protracted screening programme. DNA-diagnosis opened, in other words, the possibility to divide a known population at risk for FAP into a carrier group which can be followed with traditional clinical screening methods, and a non-carrier group which may be excluded from risk and relieved from participation in a screening programme. For those appearing to be carriers, DNA-diagnosis may have additional value in decisions about prophylactic interventions, and is available in the form of prenatal diagnosis.[2] Thus, options for DNA-diagnosis were readily incorporated in clinical practice as a diagnostic tool which contributes to more efficient and improved forms of preventive care.

A new regime of prevention

Familiar Adenomatous Polyposis (FAP), over the course of time, transformed from a 'cancer in the colon that happens to run in families'[3] to a disease 'caused by a mutation in the APC locus on chromosome 5q21'. With this shift in definition, other things changed as well. In our history of FAP, we have come across different situations where patients and their families, doctors and other health care workers met, where questions were posed, examinations took place, decisions were taken and choices were made. Our story moved not only in time, but also to different places: from the clinic and consultation room to the patient and family at home, to the clinical genetics centre. What we have tried to do is to trace the changing configurations of artefacts, practices, professions, users and institutions in which a particular case of 'new genetics' gradually has been taking shape.

We have described these changing configurations as an extending *network*, which involved new actors – family members potentially at risk, social workers, a national registry, clinical geneticists – and which established new alignments between these actors. In this network, the provision of information to family members about the hereditary nature of the disease, the establishment of family-trees, the collection of medical data, the offer of DNA-diagnosis and clinical screening, became standard to the work of the medical specialist, social worker or clinical geneticist. The data collected through this (net)work, by the Foundation for the Detection of Hereditary Tumours, facilitated not only the organisation of a national screening programme, but also the systematic monitoring of its effects and the development of guidelines to be observed in practices of diagnosis and screening. Thus, the extending FAP-network embodied a long envisaged task to improve the management of polyposis as a hereditary disease.

Our previous account of the emergence of a FAP-network not only describes the mutual shaping or co-evolution in which FAP transformed from a 'fatal disease of the colon' into a 'hereditary disorder with preventable consquences'. It also shows how this process of co-evolution crystallized into a specific pattern of roles and responsibilities which for the actors involved were difficult to deny. In the 1980's, actions and interactions in the evolving FAP-network both shaped and were shaped by what we might call a new *regime of prevention.*[4] As the publications in the Dutch Journal of Medicine show, in the early eighties, medical specialists already were expected to inform FAP-patients about the hereditary nature of the disease, and the consequent implications for members of the family. The responsibility of the physician could no longer stop with the treatment of a patient, but should extend into the patient's family. Relatives had to be informed about potential risks and if necessary, according to ruling standards, considered for regular screening. With the establishment of a national registry, the responsibility of individual specialists to offer information and care to the family of patients became institutionalised on a more collective level in the working practices, data-base and protocols of the Foundation for the Detection of Hereditary Tumours. In other words, through the efforts of the Foundation, practices of early detection and prevention no longer depended primarily on local initiatives, but had become part of a larger cancergenetic regime of prevention.

The introduction of DNA-diagnosis: a regime of self-determination

In the 1990s, the organisation and management of FAP underwent further change. Established links between patients, at-risk families, medical specialists and the Foundation were transformed both in content and in character due to the advent of DNA-diagnosis and the involvement of clinical geneticists. Through its well-kept database the Foundation had become a major resource for large-scale genetic research. As the director of the Foundation has repeatedly pointed out in many publications, the aim of the Foundation was twofold: patient care and scientific research (Vasen *et al.* 1988, 1991, 1993 and 1999). The establishment of a national FAP-registry thus was a major impulse for research on FAP and in particular research on genetic markers as was done at the molecular laboratory of the University of Leyden (Griffioen *et al.* 1999).

When, in the early nineties, researchers in Leyden found genetic markers on both sides of the so-called APC (Adenomatous Polyposis Coli) gene, it became possible to identify gene carriers in at-risk families through linkage studies (no direct mutation analysis was possible yet). The provision of a DNA-test, however, did not come within the province of the gastroenterologist or the Foundation for the Detection of Hereditary Tumours. It was the Leyden Centre for Clinical Genetics that organised and facilitated the introduction of DNA-diagnosis in FAP-families. Historically, centres for clinical genetics in the Netherlands have a privileged position in offering genetic counselling and genetic testing, and function as 'gatekeepers' for those seeking genetic consultation and diagnosis (Nelis, 2000). In these centres, genetic diagnosis is embedded in a practice of counselling in which autonomous decision-making of patients and individuals at risk is the guiding principle. When providing information, counsellors consider it as their task to be neutral and non-directive.[5] The responsibility for decisions and actions to be taken is delegated primarily to the individual asking for information and advice. In the practice of genetic counselling the principle of informed decision-making is also upheld by the relatively long time available for each consult, the obligatory time-frames between consults when clients have to make important decisions, and the extensive documentation of consultations that counsellors provide their clients with.

In other words, when DNA-diagnosis became available for those at risk for FAP, it was embedded in a practice of clinical genetics that already constituted a regime of its own. In this regime, *self-determination* was the guiding principle that defined the roles and responsibilities of the actors involved (Nelis, 1998 and 1999). The history of this clinical genetic regime is strongly related with the advent of prenatal diagnosis and the option of selective abortion, which made autonomous and informed decision-making by individual parents a particularly sensitive issue. In the Netherlands, this clinical genetic regime served as a 'niche' in which, in the 1980s, the first new options for DNA-diagnosis further shaped the practice of clinical genetics as a regime of self-determination, insofar as these options primarily brought along new opportunities for (reproductive) choice.

In this context it is interesting to see how medical specialists, in the early 1990s, considered the prospects of DNA-diagnosis in the field of cancergenetics. They referred, first of all, to the promise of improved forms of preventive care, but also pointed out that genetic counselling would deserve particular attention because those at risk for (rare) hereditary tumours would have to face more complex choices (Vasen and Müller, 1991). This comment, no doubt, strongly reflected the history and position of clinical genetics as a practice in which DNA-diagnosis was made available to patients and individuals at risk primarily as an opportunity of (informed) choice. In other words, in our story of the emerging FAP-network, the introduction of options for DNA-diagnosis did not only involve the extension of this network with a few new elements, but also created a situation in which two different regimes began to interfere with each other. Indeed, from the 1990s, FAP patients and individuals at risk found themselves being addressed in different ways: as subjects who need preventive care and as subjects who have to deal with (new) opportunities of choice.

Interfering regimes

To see what happened when a (clinical genetic) regime of self-determination began to interfere with a (cancergenetic) regime of prevention, we may take up again our description of the FAP-network and look how choices, roles and responsibilities take shape in different locations. In the following, we will move from the medical specialist at the clinic and consultation room, to the social worker visiting a family at home, to the counsellor at the clinical genetics centre.[6]

The gastro-enterologist

The following comments of a gastro-enterologist about his experiences with FAP-patients show how choices, roles and responsibilities are defined in the consultation room:

> The days that the doctor knew best are well over and done. Choices and risks related to the timing and nature of surgical interventions have all to be clearly discussed with the patient. There are a lot of things that have to be considered, including of course the wishes of the patient. [...] Of course, when a colon is full of polyps, well then it is our task to tell the message, to tell what must be done, and that is, you know, what always will be done.

As medical specialists have been more and more convinced of the necessity to regularly screen members of families at risk of FAP, they have created for patients and their relatives new courses of action and thus also new opportunities of choice. When patients are asked to inform other members of the family, they have to make a choice and may face difficult decisions, especially when relations in the family are disturbed. When family members are informed about being at risk, they may decide to undergo regular examinations, but may also prefer to keep aloof. However, in creating these courses of action, medical specialists are not seeking more room for choice, but are seeking opportunities to improve care. Indeed, diagnosis, treatment and screening of FAP are seen as matters of (early) death or (longer) life and from this perspective decisions about preventive measures – for example in relation to screening intervals – are primarily perceived as medical issues (Menko *et al.* 1999). Such decisions thus will be generally discussed and presented in terms of necessary interventions about which there is little to choose.

The social worker

The same points can be made in regard to the activities undertaken by the Foundation for the Detection of Hereditary Tumours. On the one hand, workers of the Foundation emphasize that they are allowed to approach and register individuals only on the basis of (written) consent and thus at some points in time they explicitly single out their clients as subjects of choice. Yet, the primary aim of the Foundation is to guarantee optimal care and so it does everything in its power to encourage individuals to cooperate. A social worker explains:

> Because of privacy regulations we are not allowed to approach members of the family without their personal consent. Thus patients are invited by the foundation to inform relatives at risk. Personal contact with patients at home makes it easier to persuade them that it is necessary to inform other members of the family and also to convince them of the benefit of screening. Sometimes, when patients are reluctant, the family doctor is called in. If members of the family don't want to be informed, then there is nothing more to be done.

Of course, even in this situation individuals will make different choices. Sometimes a patient is not prepared to inform other members of the family. And not everyone at risk really wants to be informed. In living their lives and their disease, people thus may choose to follow different trajectories. However, with the emergence of a practice of clinical screening and its development into a cancergenetic regime, some trajectories have been made more comfortable and predictable than others. In the early eighties it required a great deal of effort for a medical specialist to maintain a programme of screening that would allow every individual at risk to be informed and undergo regular examination. Today, a national registry, social workers who visit patients at home, information leaflets, a system of recall, working groups, survival rates, guidelines, form the constitutive elements of a regime of prevention, offering patients and relatives at risk a trajectory which is difficult for them to refuse.

The genetic counsellor

In addition to the clinic and the registry, clinical genetic centres recently also have become part of the trajectory that FAP patients and relatives at risk are supposed to follow. In the context of the established cancergenetic regime for FAP, the introduction of DNA-diagnosis is generally seen as an opportunity to improve the organisation of screening. However, in order to be referred to a clinical genetics centre, a patient or individual at risk for FAP is asked to sign a written consent and then will move along a trajectory structured by a regime of self-determination. Indeed, when an individual at risk of polyposis consults a genetic centre for DNA-diagnosis, the counsellor will meet this person on the basis of a specific protocol. In the words of a counsellor:

> According to the protocol, applicants for a pre-symptomatic test first of all will see a clinical geneticist and a psychologist. The clinical geneticist discusses the history of the disease and the personal reasons for a pre-symptomatic test. Then there is a meeting with one of the psychologists who discusses the implications of pre-symptomatic testing. This is followed by a four-week period to think the matter over, after which the applicant returns for a final discussion and decision about the test. The protocol applies to all forms of presymptomatic testing, yet things are different with polyposis of course. With FAP the consequences are very straightforward inded. [...] Thus, counsellors may often deviate from the protocol. It really depends on the circumstances. [...] In keeping track of families however, clinical genetics is more reticent and less directive than the foundation (for the detection of hereditary tumours). [...] As geneticists, we sometimes perhaps have a bit of a holier-than-thou attitude.

Thus we see how the interplay of forces actors have to deal with in the FAP-network is different in different locations and, on some points, clearly reveals an interference between two different regimes. In a regime of *prevention*, everything is done in order to promote the health of patients and relatives at risk. In this context the notion of choice only appears as a valuable opportunity or as a boundary that one should respect (the time is past in which the doctor knew best; with presymptomatic testing individuals may have to face complex choices).[7] In a regime of *self-determination* everything is done to uphold non-directiveness and informed freedom of choice, although one need not always have a 'holier-than-thou' attitude (choices may be straightforward; one may deviate from the protocol).

Conclusion: the new genetics in society

What does our case-study tell us about the new genetics in society? In discussions about the implications of the introduction of new genetic tests, attention often is focussed on the complications and ambivalences of the individual choices arising from these tests. Obviously, this is an important aspect of the new genetics, constituting what we have called a regime of self-determination. In this regime, the definition and distribution of roles and responsibilities are primarily shaped by (and also shape) discrete *moments of choice* in which the individual client is called upon to decide. However, when talking about the new genetics in society, the complications and ambivalences of individual choices are only part of the story. Another conspicuous aspect of the new genetics is the incorporation of pre-symptomatic DNA-diagnosis in more comprehensive and established screening practices, constituting what we have called a regime of prevention. In such a regime, we find another definition and distribution of roles and responsibilities, shaping a particular *trajectory* that patients and individuals at risk are invited or even urged to follow in regard to their own future health and survival.

The notion of different regimes makes clear that 'the' new genetics does not refer to a single development, practice or set of questions. It incorporates a range of different and evolving promises, opportunities, roles, choices and responsibilities that may relate to each other in complex and ambiguous ways. In our history of the FAP-network we have tried to understand the complexity and ambiguity of the new genetics in terms of two interfering regimes. In this respect, however, the story of FAP does not stand on its own. Elsewhere, we have described a similar pattern in the history of screening and DNA testing for hereditary breast cancer in the Netherlands (Nelis and Diergaarde, 1997; Nelis, 1999; Bourret *et al.* 1998). In this case too, we see two different regimes interfering in a new and emerging practice of predictive screening and testing. On the one hand, we find clinical geneticists firmly committed to non-directiveness and neutrality in regard to decision-making of women about predictive testing. On the other hand, we find medical specialists traditionally involved in the treatment and cure of individual patients and more recently also with screening and preventive surgery of women who are identified as being at risk. As pre-symptomatic DNA-diagnosis becomes more and more a means to identify those individuals at risk who may benefit from a preventive treatment – based on evidence that medically speaking a *best* option may exist – it will more and more challenge the focus on neutrality and non-directiveness as embodied in the practice of clinical genetics.[8] Thus, as the new genetics progresses, we may expect further transformations and 'regime shifts' as a result from attempts to solve the ambiguities, ambivalences and tensions arising from the interference between the two regimes we have described in this article. An interesting example in the Netherlands is the institution of so-called 'multidisciplinary clinics for hereditary tumours' in which specialists from different fields are working together, including molecular geneticists, clinical geneticists, oncologists and specialists involved in psycho-social care (Menko *et al.* 1999). These clinics thus incorporate on an institutional level aspects of both a regime of self-determination and a regime of prevention.

Notes

1. We would like to thank Brenda Diergaarde for her contribution to the research that was performed in order to write this chapter.
2. In practice, however, the option of prenatal diagnosis appears to be rarely used (Whitelaw *et al.* 1996).
3. In the late nineteenth, early twentieth century FAP used to be called 'polyposis intestini'. It was only later that the term FAP has been introduced to define the genetic origin of the disease (Palladino, 2002).
4. According to Rip and Kemp (1998) a technological regime may be defined as "the rule-set or grammar embedded in a complex of engineering practices, production process technologies, product characteristics, skills and procedures, ways of handling relevant artefacts and persons, ways of defining problems – all of them embedded in institutions and infrastructure" (p.338).
5. This does not imply that other medical practices are naturally directive. What we argue here is the particular emphasis on patient autonomy in the definition and practice of clinical genetics. Whether a non-directive approach is actually possible is another matter. For a critical review of non-directiveness and neutrality see Van Zuuren (1996, 1997), Steendam (1996) and Michie *et al.* (1997).
6. Our observations are based on interviews with a gastro-enterologist from the University Hospital in Nijmegen, a social worker from the Foundation for the Detection of Hereditary Tumours, and a clinical geneticist from the Centre of Clinical Genetics in Leyden.
7. Thus, in the literature one finds examples of a rather firm decision-making attitude on the side of the doctor while at the same paying attention to the fact

such decisions have to be seriously discussed with and considered by their patients. For example, in a recent article by Griffioen et al (1999) it is stated on the one hand that screening of first-degree relatives of FAP-patients is *mandatory* when colorectal cancer is found in the family at a young age and that a total proctocolectomy is not *recommended* by most surgeons, while at the same time the authors acknowledge that "the physician (...) should be aware of these and other factors and should discuss them thoroughly with the patient" (p.126).

8. In this context, the notion of non-directiveness and neutrality is also questioned by genetic counsellors themselves. See for example Tibben, 2002.

References

Annual Report (1994), Netherlands Foundation for the Detection of Hereditary Tumours, Leyden, University Hospital.

Bourret, P., Koch, L. and Stemerding, D. (1998), 'DNA diagnosis and the emergence of cancer-genetic services in European health care' in P. Wheale, R. von Schomberg and P. Glasner, *The Social Management of Genetic Engineering*, London: Ashgate, pp. 117-138.

Griffioen, G., Vasen, H.F.A., Verspaget, H.W. and Lamers, C.B.H.W. (1999), 'Familial adenomatous polyposis: from bedside to bench and vice versa', *Cytogenetics and Cell genetics*, 86, pp. 125-129.

Koch, L. and Stemerding, D. (1994) 'The sociology of entrenchment: a cystic fibrosis test for everyone?', *Social Science and Medicine*, 39, pp. 1211-1220.

Menko, F.H., Griffioen, G., Wijnen, J.TH., Tops, C.M.J., Foddy, R. and Vasen, H.F.A. (1999), 'Genetica van darmkanker. I. Non-polyposis en polyposisvormen van erfelijke darmkanker', *Nederlands Tijdschrift voor Geneeskunde*, 143, pp. 1201-1206.

Menko, F.H., Griffioen, G., Wijnen, J.TH., Tops, C.M.J., Foddy, R. and Vasen, H.F.A. (1999), 'Genetica van darmkanker. II. Erfelijke achtergrond van sporadische en familiaire darmkanker', *Nederlands Tijdschrift voor Geneeskunde*, 143, pp. 1207-1211.

Michie, S., Bron, F., Bobrow, M. and Marteau, T.M. (1997), 'Non-directiveness in Genetic Counselling: an Empirical Study', *American Journal of Humam Genetics* 60, pp. 40-47.

Nelis, A. (1999) 'Managing genetic testing: the relative powerlessness of actors in stable practices', *New Genetics and Society*, 18, No. 2/3, pp. 125-143.

Nelis, A.P. (2000) 'Genetics and Uncertainty', in N. Brown and B. Rappert (eds) *Contested Futures*, Ashgate Publishers, pp. 209-227.

Nelis, A.P. and Diergaarde, B. (1997), 'Erfelijke Borstkanker – De Keuzes', *Tijdschrift voor Gezondheid and Politiek*, no. 3, pp. 24-26.

Palladino, P. (2002) 'Between Knowledge and Practice: On medical professionals, patients and the making of the genetics of cancer', *Social Studies of Science*, 32/1, pp. 137-165.

Rip, A. and Kemp, R. (1998) 'Towards a Theory of Socio-Technical Change', in S. Rayner and E.L. Malone (eds) *Human choice and Climate Change* (volume II), Columbus, Ohio: Battelle Press, pp. 329-401.

Ruiter, P. de, den Hartog Jager, F.C.A. (1981), 'Polyposis coli', *Nederlands Tijdschrift voor Geneeskunde* 125, pp. 1739-1744.

Slooten, E.A. van (1981), 'Polyposis coli', *Nederlands Tijdschrift voor Geneeskunde* 125, pp. 1762-1763.

Steendam, Guido (1996) *Hoe genetica kan helpen*, Acco: Leuven.

Tibben, A. (2002) Genomics and Dissemination of Genetic information to individuals at risk: the need for a proactive approach? *NWO essay*. (http://www.nwo.nl/NWOHome.nsf/pages/NWOP_5DPCTM?Open Document).

Vasen, H.F.A. and Müller, H.J. (1991), 'DNA-onderzoek in families met erfelijke vormen van kanker', *Nederlands Tijdschrift voor Geneeskunde*, 135, pp. 1620-1623.

Vasen, H.F.A., Griffioen, G., Lips, C.J.M. and van Slooten, E.A. (1988), 'De waarde van screening en landelijke registratie van erfelijke tumoren', *Nederlands Tijdschrift voor Geneeskunde*, 132, pp. 1609-1612.

Vasen, H.F.A., Nagengast, F.M., Griffioen, F., Kleibeuker, J.H., Menko, F.H. and Taal, B.G. (1999) 'Periodiek colonscopisch onderzoek bij personen met een belaste familieanamnese voor colorectaal carcinoom', *Nederlands Tijdschrift voor Geneeskunde*, 143, pp. 1211-1214.

Vasen, H.F.A. and Vermey, A. (1993), 'Nieuwe ontwikkelingen in onderzoek van erfelijke kanker', *Tijdschrift Kanker*, 2, pp. 54-57.

Whitelaw, S., Northover, J.M. and Hodgson, S.V. (1996) 'Attitudes to predictive DNA testing in familial adenomatous polyposis', *Journal of Medical Genetics*, 33, pp. 540-543.

Zuuren, F.J. van (1996) 'Neutraliteit in de praktijk van de genetische counseling', *Tijdschrift voor Geneeskunde en Ethiek* 6 (3), pp. 77-81.

Zuuren, F.J. van (1997), 'The standard of neutrality during genetic counselling: An empirical investigation', *Patient Education and Counselling* 32, pp. 69-79.

12 Lumping and Splitting Revisited: Or What Happens When the New Genetics Meets Disease Classification[1]

Adam M. Hedgecoe

A number of authors have suggested that the rise of the new genetics will allow for more accurate disease classification and better diagnosis of even common conditions. For example, John Bell writes that: 'Perhaps the single most important contribution of the new genetics to health care is that it will create a biological ... framework with which to categorise diseases'. To be specific, the effect of the new genetics will be

> the subdivision of heterogeneous diseases ... into discrete entities ... Understanding the biological evens and pathways identified by genetics as contributing to disease will lead to clear definition of disease.
>
> (Bell 1998, 618-619. See also Cookson, 1999)

In this chapter I explore the way in which the new genetics impacts on disease classification and suggest that Bell is only partly right. Genetics can lead to more refined classification and subdivision but it can also lead to confusion and ambiguity. Problems arise when we assume that genetic information can provide us with a clear definition of what a disease is, free from clinical expertise. Bowker and Star (1999) have shown how disease classifications (and other categories) are constructed, and the way in which they have to balance the tension between competing social interests. What comes across in both of the examples that I present here is the way in which genetic research alters disease classifications in ways which are not necessarily helpful to the clinicians treating patients.

The title of this chapter is inspired by the eminent medical geneticist Victor McKusick who founded *Mendelian Inheritance in Man* (now *Online Mendelian Inheritance in Man*), the standard reference work for medical genetics. In his paper 'On Lumpers and Splitters, or the Nosology of Genetic Disease' (1969) McKusick suggests that the medical geneticist must span the two traditional categories among those involved in classification. Those who pull different features together as evidence of one underlying factor (lumpers) and those who subdivide classes of objects into smaller sub-groups (splitters). This chapter takes these ideas and shows how genetic information leads to lumping and splitting in disease classification.

Rather than simply preach at scientists and clinicians and point out how inappropriate their ideas about genetic disease classification are (Hull 1978; Hesslow 1984; Wachbroit 1994), this chapter examines the way in which disease classifications actually change as a result of genetic information. To do this, I will use two example diseases which have undergone reclassification as a result of genetic information: Cystic Fibrosis and Diabetes.

Cystic fibrosis

It may seem strange to suggest that cystic fibrosis disease classification changed as a result of genetic information; cystic fibrosis has been seen as running in families since 1952, although it was first clearly recognised as a distinct condition in the long hot New York summer of 1938 (Super, 1992; Mearns, 1993). So closely has Cystic Fibrosis (CF) been associated with genetic diseases that it has become for many the classic, monogenic recessive disorder. It is caused by mutations in a single gene, and one needs to inherit two copies of the faulty gene (one from each parent) to develop CF. In many ways Cystic Fibrosis has become the 'poster boy' for the new genetics. In the early stages of the Human Genome Project it was often cited as a good example of the kind of condition that the HGP would solve (OTA, 1988; Jones, 1993; Bodmer and McKie, 1994). It became an early target for gene therapy experiments (Goldspiel, Green and Calis, 1993).

Most importantly for my purposes, it became the first disease to have its gene identified using a technique called 'reverse genetics', proposed as the way in which information from the HGP would be translated into therapeutic benefits for patients. Reverse genetics allows researchers to identify a specific gene and then 'work forward' to identify the gene product. In the case of Cystic Fibrosis, the gene in question led researchers to a protein, named Cystic Fibrosis Conductance Regulator (CFTR), which is involved in transport of sodium and chloride ions across cell membranes (Iannuzzi and Collins, 1990).

In this sense, CF turned from an inherited disease which was known to run in families, to a genetic one, where the specific gene involved was identified. The identification of the CF gene and the protein that it codes for had a significant impact on the way in which clinical researchers viewed the condition Cystic Fibrosis. Scientists began to identify different mutations in the CFTR gene, and to work out how they effected the protein's structure and function to bring about the various symptoms of CF. This in turn began to feed into the clinical debates surrounding what exactly *counts* as cystic fibrosis.

Classification of CF prior to 1989

Prior to the discovery of the gene that codes for CFTR, the classification of cystic fibrosis was based on clinical symptoms, with help from some laboratory-based tests. The most obvious symptom of CF is the chronic lung infections and breathing difficulties that arise from large amounts of mucus building up in the lungs. Other common symptoms are low weight (through poor nutrient absorption) and high levels of sodium in the sweat (Hodson, 1984).

This last feature was used to apply the pre-genetic 'gold standard' test for CF, the sweat test. In this procedure, an electrical current is used to drive a chemical (pilocarpine nitrate) into the skin which in turn stimulates the production of sweat which is collected on a gauze. The levels of chloride in the sweat then suggest the presence or absence of CF (Rosenstein, 1990; Birnkrant and Stern, 1991).[2]

Table 12.1 Clinical symptoms of cystic fibrosis

Chronic sinopulmonary disease manifested by: Persistent colonisation/infection of the lungs with typical CF pathogens Chronic cough and sputum production Persistent chest X-ray abnormalities Airway obstruction manifested by wheezing and air trapping
Gastrointestinal and nutritional abnormalities including: Intestinal: rectal prolapse, intestinal obstruction syndrome Pancreatic: pancreatic insufficiency, recurrent pancreatitis Hepatic: chronic hepatic disease Nutritional: failure to thrive (protein-calorie malnutrition)
Salt loss syndromes: Acute salt depletion
Male urogenital abnormalities resulting in obstructive azoospermia

(adapted from Rosenstein and Zeitlin 1998: 277)

Of course there are other conditions which produce some of the same symptoms as CF, sweat chloride levels can be elevated by other diseases and the sweat test itself is complex, and needs an experienced operator to avoid false positives. But despite this, the diagnosis of CF was made by clinicians on the basis of clinical symptoms, and the disease was classified in terms of the symptoms that arose. But at the beginning of the 1990s, it was also clear that the discovery of the CF gene and the protein CFTR would have a serious impact on clinical practice, and would hopefully provide definite, unequivocal diagnoses of those patients where a diagnosis of CF was not clear; 'genetic screening may eventually supplant...[the sweat test]...for definitive diagnosis' (Birnkrant and Stern 1991: 97).

Classification of CF after 1989

The availability of genetic tests changed the classification of Cystic Fibrosis; in McKusik's terms, CF 'lumped' rather than 'split'. The boundaries of the classification expanded to include neighbouring conditions not generally regarded as Cystic Fibrosis. The first point to note is that much to everyone's surprise, the gene which codes for CFTR and which goes wrong in CF has an extraordinarily large number of mutations. The current count stands at over 1,000 different mutations, each of which would theoretically need to be tested for to rule out CF. Fortunately, most of these mutations are very rare, only being reported in a few families, with a small number of mutations being the cause of most cases of CF.[3]

In her detailed paper on this topic, Anne Kerr (2000) explores the way in which, over the 1990s, scientists constructed a nosological link between cystic fibrosis and a form of male infertility called Congenital Bilateral Absence of the Vas Deferens (CBAVD). While CBAVD has long been recognised as a symptom of Cystic Fibrosis, there are also a number of other conditions for which it was indicative. Research in the 1990s showed that there were unexpectedly high number of mutations in the gene coding for CFTR in men diagnosed as infertile with CBAVD (Dumur *et al.* 1990; Anguiano *et al.* 1992; Gervais *et al.* 1993; Culard *et al.* 1994. See Lissens and Liebaers, 1997 for review). Sometimes these men had rare mutations which were not found in typical cases of CF, sometimes they had mutations in non-coding regulatory sections of the CFTR gene and occasionally they were only heterozygous for CF mutations, i.e. they only carried one copy of the faulty gene. Nevertheless, despite the atypical features of the genotypes of many men with CBAVD there was 'an epistemological shift from considering CBAVD to be one symptom of CF in men, to a mild or genital form of CF' (Kerr, 2000: 861).

This shift did not go unnoticed in the CF scientific and medical communities, and clearly some people were unsure about such reclassification. The choice they faced was 'whether all cases of CBAVD as an isolated finding should be uniformly classified within the spectrum of CF or be defined as CFTR associated but distinguishable from clinical CF' (Colin *et al.* 1996: 442). The answer made in the CF literature is that 'CBAVD and cystic fibrosis are extreme forms of a wide nosologic spectrum of conditions that have a common molecular basis' (Chillón et al, 1995: 1479-1480).

My own work in this area mirrors Kerr's focus on CBAVD and updates the debate on CF classification (Hedgecoe, forthcoming). As research has progressed, more and more 'bordering' conditions have been linked to CF on the strength of sufferers of those conditions having higher than expected levels of CFTR mutations in their populations. As well as CBAVD and milder forms of male infertility (such as Congenital Unilateral Absence of the Vas Deferens – CUAVD), conditions such as: idiopathic pancreatitis, disseminated bronchiectasis, allergic broncho-pulmonary aspergillosis, atypical sinopulmonary disease, diffuse bronchiectasis associated with rheumatoid arthritis and sarcoidosis have all been linked to defects in the gene which codes for CFTR (Zielenski, 2000: 125). The question is can these different conditions be classed as forms of cystic fibrosis, in the way that CBAVD became a 'mild genital' form of CF?

The literature suggest a great deal of uncertainty on the part of clinicians and researchers about this. Taking part in a debate over whether idiopathic pancreatitis should be included in the spectrum of conditions classed as CF, Sharer, Swartz and Path reveal the concerns present in the cystic fibrosis medical community. Far from improving diagnosis, the discovery of the CFTR-gene has meant that 'The definition of cystic fibrosis has become progressively more hazy' (Sharer, Swartz and Path, 1999: 238). As in the case of CBAVD, there is a tension between clinical practicalities and the genetic reclassification: 'Because the phenotypic spectrum that may now legitimately be called cystic fibrosis has become so large, it is unhelpful in the clinical context' (ibid). Since 'the time is right to recognize that the diagnosis of cystic fibrosis is too nebulous to preserve in clinical practice', they suggest a new classification system is needed. This would include:

- Cystic fibrosis 'disease' which is 'the progressive suppurative respiratory tract condition irrespective of other phenotypic features' and

- Cystic fibrosis 'syndrome' which includes 'pancreatic manifestations, congenital absence of the vas deferens, and lesser pulmonary manifestations' (ibid).

Although in 1998 a panel of experts attempted to produce a consensus statement on CF diagnosis (Rosenstein and Cutting, 1998) a number of recent discoveries have challenged how useful this statement is. As a result 'We are left with the paradoxical situation of having some patients with typical CF in the absence of mutations, and others with CFTR mutations in the absence of clinical features' (Rosenstein, 2002: 84).

While it might be tempting to conclude that the debate over the role of mutation testing for CF and its effect on disease classification is an interesting, but minor part of the ethical issues surrounding genetic tests, this would be to ignore the importance of classification and what we call things. As Bowker and Star point out when things are perceived as real, then they have real consequences:

> even when people take classifications to be purely mental, or purely formal, they *also* mould their behaviour to fit those conceptions. When formal characteristics are built into wide-scale bureaucracies such as the WHO...then the compelling power of those beliefs is strengthened considerably.
>
> (Bowker and Star, 1999: 53)

Consequences of CF 'lumping'

Reclassifying people effects their lives. To tell a man who has thought of himself as infertile that he 'really' has a serious genetic disease which usually kills people before they reach 35 is obviously going to have an impact on his life. It does make sense to genetically test men with CBAVD since they may be at increased risk of developing other CF symptoms (such as lung problems). Also, since new technology means that men with CBAVD can now have children, there is also a need for genetic testing and counselling for CBAVD men and their partners who undergo this kind of treatment; they may be at increased risk of having a child with 'full' CF, and it would be irresponsible of their doctors not to provide them with the best possible information (Sawyer, Tully and Colin, 2000).

But at the same time, there seems to be reluctance on the part of these men to be seen as CF sufferers. Colin *et al.* note that only 18 of the 50 CBAVD men in their study 'reclassified' as having CF accepted the offer of free and confidential follow up evaluation. These authors suggest that:

> The diagnosis of male infertility appears to be sufficiently stressful for many men that the possibility of making an additional diagnosis, especially one carrying the social stigmatization of CF may be avoided...perhaps, it may be that the concept of having an underlying chronic illness is inconsistent with their personal belief that they are fit and healthy.
>
> (Colin *et al.* 1996: 444)

Although current diagnostic criteria explicitly state that CF diagnosis is a *clinical* procedure, the availability of genetic tests effect the way this is done. In 1999, Chmiel *et al.* (1999) reported the case of a girl who initially presented at their CF centre, aged 2 months, for a second opinion. Some of her relations had been diagnosed as CF-sufferers and both her parents were CF carriers. Since they refused prenatal testing on the foetus, it was at birth that they found that the girl had inherited both faulty versions of the CF gene and she was diagnosed as having the condition. Despite this diagnosis of cystic fibrosis, the girl remained healthy; her sweat was collected and found to be normal. Pancreatic enzyme treatment which had begun when the original diagnosis was made wad discontinued three months later with no apparent negative effect on her digestion.

The authors point out that although 'This child was given the diagnosis of CF based solely on the presence of two mutant alleles...she does not yet meet diagnostic criteria for CF' since she displays none of the clinical symptoms traditionally associated with the disease (Chmiel *et al.* 1999: 825). The reason for this apparent discrepancy is that despite having two copies of CF causing alleles, she probably *also* carries normally neutral alleles which serve to ameliorate the effect of the deleterious genes. As a result, her body produces enough working CFTR protein and she has a normal phenotype. As these authors point out, 'Had this child presented before the discovery of the CF gene, a sweat test may have been obtained...[on the basis of family history]...but the CF diagnosis would not have been made' (Chmiel *et al.* 1999: 825).

One might think that no harm had been done by the CF diagnosis; the girl is too young to understand what has happened and no harmful treatment had been carried out. Yet this would be to ignore the wider, familial effects that diagnosis of a condition such as CF has: 'The patient's father, a career serviceman in the US Navy, accepted a discharge so that his daughter could receive consistent medical care at one CF centre. Because he was then unable to obtain employment for six months, the family was forced to seek temporary housing and governmental assistance' (Chmiel *et al.* 1999: 825).

Thus in the case of cystic fibrosis, the disease where genetic testing was meant to clarify disease classification, the opposite seems to be true. This is important to bear in mind when we consider that a the end of April 2001, Yvette Cooper the UK Minister for public health announced a new initiative to expand screening for Cystic Fibrosis (CF) to every new-born in the country: 'It is important that every child gets the best start in life...evidence shows that screening newborn babies for...cystic fibrosis can make a big difference to the treatment and support they receive, and can significantly improve their health and development' (DoH, 2001).

My aim is not to suggest that there is no therapeutic value to be gained from the introduction of CF screening but simply to point out that in the case of CF, genetic testing has had an unexpected effect on the way in which the disease is classified. First it contributed to the expansion of the CF classification to incorporate the form of infertility CBAVD. Then it increased uncertainty about what counts as CF, since there were so many different conditions linked to an increased number of CFTR mutations. Despite deliberate attempts by clinicians and organisations to maintain a clinical approach to CF diagnosis, the genetic test has crept into CF classification and is impacting upon diagnosis.

Diabetes

The other case of disease re-classification I want to explore is that of diabetes. While the cystic fibrosis classification has expanded due to genetic explanation, diabetes has undergone 'splitting' into numerous sub-divisions. This change in classification started in the mid-1970s and continues today. Diabetes Mellitus is an umbrella term for a number of different diseases of glucose metabolism. Broadly, this syndrome is divided into two kinds of disease: the kind that affects young, thin people, which is life threatening and requires insulin injections to treat; and the kind that affects older people, who are often obese, and who do not rely on regular insulin injections for their survival.

In the case of the first kind of diabetes, the body's immune system has destroyed the β-cells in the pancreas, which makes insulin. Therefore, when a diabetic's blood sugar rises (after a meal, for example), the body cannot produce insulin which promotes the storage of glucose as glycogen and fat. Symptoms include: thirst, increase in appetite and glucose in the urine. In its search for energy, the body begins to break down fat, producing ketones, a process called ketoacidosis. The build up of ketones in the body can lead to coma and death. Regular injections of insulin allow the body to absorb glucose and the diabetic to survive. In the second kind of diabetes, the body's cells become insulin resistant. Therefore although the body is producing insulin, it cannot produce enough to fully overcome the insulin resistance. Symptoms include: thirst, increased urination, the slow healing of wounds, weakness, fatigue and blurred vision. It can normally be controlled with dietary changes and lifestyle alterations (Smallwood, 1990).

The condition we currently recognise as diabetes mellitus (DM) has been medically described for over 2000 years, including the recognition that it involves two distinct 'types', one 'quick and thin' and the other 'fat and slow'. In 1875 Bouchart suggested that the terms 'diabète gras' and 'diabète maigre' be used to describe two diseases with different prognoses and classifications. These ideas lapsed in the early twentieth century, as DM was described as a single disease consisting of different stages such as 'brittle'/'stable' or 'Juvenile'/'maturity'.

The idea of diabetes as a syndrome was emphasised in the late 1940s by, among others, Sir Harold Himsworth, who suggested that it be divided into 'insulin sensitive' and 'insulin insensitive' varieties. Soon after this, R.D. Lawrence proposed that a distinction should be drawn between versions of diabetes mellitus on various grounds such as: primary pancreatic destruction; disturbance of other endocrines; disturbance of fat storage; diabetes associated with obesity without ketosis; and insulin deficient diabetes. He pointed out that the different types seemed to run in different families, leading to a clear clinical division in patients: those who require insulin to survive and those who do not (Lawrence, 1951: 375). But despite these efforts, the division of diabetes into different sub-types, obvious as it is to us now, was not so clear to the medical establishment of the early 1970s 'which pictured early – and late – onset diabetes as gradations or variants of the same genetically determined disease process' (Gale, 2001: 217).

Reclassification in the 1970s

In 1976, a British clinician researcher called Andrew Cudworth revived and popularised Hugh-Jones' 1955 distinction between Type I and Type II DM (Bennett, 1985; Keen, 1986; Gale, 2001). Cudworth was one of a number of researchers investigating the role of the immune system in diabetes causation focusing on the Histo-Compatibility System. The HCS is a section of the short arm of chromosome 21 and its genes code for Human Leukocyte Antigens, proteins that sit on the surface of cells involved in the immune system. From the late 1960s onwards, it was noted that certain combinations of these antigens seemed to have associations with certain disorders, such as Hodgkins' disease. Studies such as Singal and Blajchman's (1972) suggested that there is also a relationship with diabetes.

What Cudworth did in a couple of review articles (Cudworth, 1976, 1978) was put forward a new classification system for diabetes, based on the involvement of HLA genes. He constructs a classification system for diabetes which fixes the divisions between the two kinds of diabetes on genetic grounds, and provide the impetus for the splitting of diabetes into sub-diseases (Hedgecoe, 2002). This came at a time when there was great concern in the diabetes medical community about the classification of the disease:

> Types of diabetes were loosely divided into 'juvenile onset' and 'maturity onset', with secondary diabetes, chemical diabetes, borderline diabetes and prediabetes all used in ill-defined ways.
>
> (Alberti and Zimmet, 1998a: 535)

Cudworth's ideas, and those of like-minded contributors (e.g. Irvine 1977) fed into the system devised by the National Diabetes Data Group of the American Diabetic Association, which met in 1978. In 1979, it published its definitive recommendations for diabetes diagnosis and classification. This system was drawn up with the intention of producing clear, mutually exclusive divisions on the basis of simple clinical observations:

Table 12.2 NDDG system of diagnosis and classification

- Insulin Dependent Diabetes Mellitus (IDDM): Type I
- Non Insulin Dependent Diabetes Mellitus (IDDM): Type II
- Other Types: from pancreatic disease
- Impaired Glucose Tolerance (IGT)
  - non-obese
  - obese
  - IGT associated with certain conditions/syndromes
- Gestational diabetes

(Based on NDDG, 1979)

This classification system was adopted and recommended by the World Health Organisation in 1980. Although Cudworth's focus on the HLA system is obscured in this classification system (though discussed in the notes accompanying it), it did have the effect of formally reifying the difference between two different types of diabetes, IDDM (associated with the HLA genes) and NIDDM.

Following the publication of the NDDG's classification system, debate continued over the use of both clinical (IDDM/NIDDM) and aetiological (type I/II) terms in the same classification system. Commentators suggested that it made little sense to present these two ways of viewing diabetes as synonymous, since although they overlapped, there would be cases covered by one definition that would be excluded by the other (Keen, 1985, 1986; Bennett, 1985).

Diabetes classification in the 1990s

By the 1990s, the time had come for another reorganisation of the diabetes classification system. The American Diabetes Association (ADA) convened another committee to revise diabetes nomenclature. The result is a system which defines 'Type I' diabetes as:

> Immune-mediated diabetes...[which]...results from a cellular-mediated autoimmune descruction of the β-cells of the pancreas...the disease has strong HLA associations, with linkage to the DQA and B genes, and it is influenced by the DRB genes.
>
> (ADA, 1997: 1186)

By linking to specific HLA genes, the ADA classification formalises the genetic sub-division of diabetes. It specifically eliminates the use of IDDM/NIDDM as classificatory categories, stating that: 'These terms have been confusing and have frequently resulted in classifying the patient based on treatment rather than etiology' (ADA, 1997: 1984).

Table 12.3 Etiologic classification of diabetes mellitus

I.	Type 1 diabetes (β-cell destruction, usually leading to absolute insulin deficiency)
A.	Immune mediated
B.	Idiopathic
II.	Type 2 diabetes (may range from predominantly insulin resistance with relative insulin deficiency to a predominantly secretory defect with insulin resistance)
III.	Other specific types
A.	Genetic defects of β-cell function
1.	Chromosome 12, NHF-1α (formerly MODY3)
2.	Chromosome 7, glucokinase (formerly MODY2)
3.	Chromosome 20, HNF-4α (formerly MODY1)
4.	Mitochondrial DNA
5.	Others
[seven more subdivisions (B-H), each containing between 3 and 11 subclasses of diabetes]	
IV.	Gestational diabetes mellitus (GDM).

From ADA 1997: 1185

If the use of genetic explanation 'lumped' Cystic Fibrosis, then in diabetes, genetics is a 'splitter'. There has been an explosion of different subtypes of diabetes: in 1979 there were nine types of diabetes (types I and II of various kinds), seven kinds of Impaired Glucose Intolerance (IGT), and Gestational diabetes, a total of seventeen. In 1997, the classification included 57 different types, with IGT a clinical 'stage' of the disease development rather than a category itself.

Implications of splitting

Clearly important issues arise from the reclassification of diabetes. The sub-division that results can be seen as a positive step for patients: 'better subclassification will lead to more precise targeting of specific treatments and eventually to better outcomes' (Wareham and O'Rahilly 1998: 359). And unlike CF, we might have reason to believe that effect of genetic explanation on diabetes classification will lead to clinical benefits; the splitting of classification into sub-categories avoids the broad, sweeping generalisations that resulted from the 'lumping' of cystic fibrosis. Further genetic sub-division in diabetes seems likely; it is encouraged by both genetic research and the discursive history of its classification. The idea that diabetes should be sub-divided is not questioned since at the broadest level, the subdivision of diabetes can only be positive; to blur the distinction between type 1 diabetes and type 2, would be to deny some people the insulin they (probably) need to live.

But issues do surround *when* a diagnosis is made; genetic explanations imply that diagnosis will eventually be made on genetic grounds, when patients are still asymptomatic. There is awareness that, certainly in the case of type 2 diabetes, asymptomatic diagnosis and treatment may not be the ideal option (Goyder and Irwing, 1998). A recent report on the disease haemochromatosis suggests that patients' views of medical intervention based on asymptomatic genetic diagnosis are not necessarily positive (Seamark and Hutchinson, 2000). It is not that such interventions are a bad idea, but it is notable that they should be considered as an automatic application of genetic testing for even 'simple', Mendelian diseases. How much more difficult are the choices to be made in the case of a disease like diabetes, which even if one thinks is a genetic condition, is caused by a number of separate genes? Yet the long preclinical period in Type 1 diabetes, where the real damage is done to the pancreas, is the time where interventions may be the most effective:

> Type I diabetes is an autoimmune disease with a long preclinical course [ref], the identification of individuals prior to the onset of the disease process provides a real opportunity for predictive testing and for therapeutic intervention.
>
> (Friday, Trucco and Pietropaolo, 1999: 11)

These issues become even more pressing if we consider the recent announcement from the University of Florida that it is planning a genetic screening programme to find infants at risk of developing diabetes later in life (University of Florida, 2002).

Perhaps we need to explore possible problems with such a programme by looking at the rare dominantly inherited version of type 2 diabetes, known as Mature Onset Diabetes of the Young or MODY. This condition:

> is characterized by an age of onset of less than 25 years, the correction of fasting hyperglycaemia without insulin for at least 2 years following diagnosis, nonketoic disease [i.e. no breakdown of proteins and production of ketones], and an autosomal dominant mode of inheritance.
>
> (Winter, Nakamura and House, 1999: 765)

The only in-depth psychosocial research on testing for MODY suggests that screening programmes need to be carried out with extreme levels of sensitivity and care (Shepherd et al, 2000). The authors of this study interviewed an affected family, the genetics specialist and paediatrician involved in treating them. They concluded that although:

> each request for genetic testing should be considered individually, we would suggest exploring the four main themes that arose with this family in consultations with other MODY families considering predictive genetic testing.
>
> (Shepherd *et al.* 2000: 257)

The four themes they identified were: the role of autobiographical experience (since many families will have experience with diabetes and this will alter their attitude towards testing[4]); the motivation for testing (often cited as the reduction of uncertainty); the competing priorities of health professionals and clients in genetic counselling (patients are often seeking a result, while counsellors are concerned with imparting information); and the differing attitudes towards testing in children (and in turn whether parents should have the final say on testing) (Shepherd *et al.* 2000: 256).

These results suggest, like much of the psycho-social research into genetic testing for complex disease, that asymptomatic testing for diabetes needs to be approached cautiously. This is important, especially since the authors of this study suggest that focusing on MODY will give insights into genetic testing for type 1 and 2 diabetes (Shepherd *et al.* 2000: 255-256). The British Diabetic Association's guidelines on genetic screening suggest that while 'Genetic testing of MODY families is thus feasible...genetic screening for identifying individuals at high risk of developing [type 1] diabetes is not helpful' (Avery *et al.* 1998).

Yet genetic distinctions between kinds of diabetes are not just being drawn within the research literature, but also within clinical discourse. A recent review for the *British Medical Journal* points out that type 1 diabetes in early childhood is associated with heterozygosity for HLA DR3/4, while its onset in older children and adolescents is linked with HLA DR3:

> Whether these groupings represent separate diseases or simply reflect the rapidity with which β cell destruction takes place is unclear.
>
> (Mandrup-Poulsen, 1998: 1221)

Conclusion

What do these two brief case studies tell us about the effect of genetic explanations on disease classification? At the simplest, and most banal, level, it tells us that genetics leads to both lumping and splitting. Partly these changes are due to the nature of the genes involved, but the role of social factors cannot be ignored. If I had chosen other conditions, haemochromatis and Alzheimer's disease for example, then the same pattern of lumping and splitting would be visible, although the time scales and social aspects involved would not necessarily be the same. The predictions of the supporters of the new genetics are not exactly on target. Genetic explanations do not necessarily lead to sub-division, nor do they automatically reduce ambiguity and confusion in disease classification.

We should also note that lumping and splitting do not have to be caused by genetic explanation. In CF, the sweat test helped incorporate pancreatic insufficiency into the spectrum of conditions classed as CF (Stern, 1998), and in diabetes, autoimmune explanations beyond the simple HLA system contributed to the sub-division of the classification system (Gale, 2001). But while the overall effect of genetic explanation may not be 'exceptional', these cases do suggest that one of the factors which drive lumping and splitting is; the relationship between clinicians and researchers.

A consistent theme through both these case studies is the tension between the needs of clinicians, and the requirements of researchers. At its most simple 'a classification appropriate to a clinician whose concern is with diagnosis and treatment may well be inappropriate to a basic scientist whose concern is research strategy and experimental design' (Keen, 1985: 31). The interest of the clinician is the individual patient, and from this point of view: 'Classification schemes and diagnostic criteria should, above all, be utilitarian. They should help physicians select a management program which is of maximum benefit and minimum harm to *individual patients* in their offices today' (Genuth, 1982: 1191).

In CF some are concerned that: 'a dichotomy is emerging between the perspective of the scientist and that of the clinician and patient. Although the scientists may consider that CBAVD is theoretically part of the spectrum of CF, this definition is not necessarily clinically helpful...we suggest caution in describing CBAVD alone as equivalent to CF in the clinical setting' (Colin *et al.* 1996: 444). In diabetes, the refuge in clinical application has obvious echoes of the debates surrounding cystic fibrosis although it is less explicit intention of improving clinical care, and making diagnosis easier for both doctors and patients. Yet some in the medical community suggest that the nature of modern diabetes research will lead to an increase in the separation between researcher and clinician:

> It is clear to me that we are unable to function as basic scientists and caring physicians at the same time, as ideal as this may have been in the past...[instead]...we must constantly try to incorporate scientific, critical thinking into clinical medicine.
>
> (Berger, 1996: 754-755)

Of course it would be over-stretching my data to claim that tensions between researchers and clinicians only occur as a result of genetic explanations. But it does focus attention on the way in which genetic explanations are promoted within medicine, and highlights the need for a wider debate about exactly what we want disease classifications to do. While these kinds of debates do occur within disease specialisations, it is rare for those beyond them to take part. Of course we need broader public debate about genetic screening programmes and the use of genetic tests by third parties (e.g. insurance companies). But it might also be useful to raise debate about 'internal' scientific issues as well, such as what counts as a disease.

Notes

1. Thanks to Jon Turney for comments on an earlier draft. This research was supported by an ESRC Postgraduate Grant, No. R00429834453.
2. Chloride levels of less than 40 mmol per litre of sweat is normal, more than 60 mmol per litre abnormal and indicative of CF and a level between 40 and 60 equivocal and indicates the need for further examination.
3. For example the mutation ΔF508 causes up to seventy percent of CF cases worldwide (Zielenski, 2000).
4. In this case, Bob (the father)'s sister had died from diabetes in her early 30s.

References

Alberti, K.G.M.M. and Zimmet, P.Z. (1998b) 'Definition, Diagnosis and Classification of Diabetes Mellitus and its Complications. Part 1: Diagnosis and Classification of Diabetes Mellitus, Provisional Report of a WHO Consultation' *Diabetic Medicine*, 15, pp.539-553.

American Diabetes Association (1997), Report of the Expert Committee on the Diagnosis and Classification of Diabetes Mellitus, *Diabetes Care*, 20, pp.1183-1197.

Anguiano *et al.* (1992) 'Congenital bilateral absence of the vas deferens. A primarily genital form of cystic fibrosis' *JAMA*, 267, pp. 1183-1197.

Avery *et al.* (1998) 'British Diabetic Association Guidelines on Genetic and Immune Screening for Type 1 Diabetes Mellitus', *Diabetic Medicine*, 15, p. 643.

Bell, J. (1998) 'The new genetics in clinical practice', *British Medical Journal*, 316, pp. 618-620.

Bennett, P.H. (1985) Basis of the Present Classification of Diabetes, pp. 17-29 of Vranic, M., Hollenberg, C.H. and Steiner, G. (eds) *Comparison of Type I and Type II Diabetes: Similarities and Dissimilarities in Etiology, Pathogenesis and Complications*, Plenum Press, New York and London.

Berger, M. (1996) 'To bridge science and patient care in diabetes', *Diabetologia*, 39, pp. 749-757.

Birnkrant, D.J. and Stern, R.C. (1991) 'Sweat Testing in the 90s', *Americal Journal of Asthma and Allergy for Pediatricians*, 4 (4), pp. 194-198.

Bodmer, W. and McKie, R. (1994) *The Book of Man. The quest to discover our genetic heritage*, Little, Brown, London.

Bowker, G.C. and Star, S.L. (1999) *Sorting Things Out: Classification and its consequences*, MIT Press, Cambridge, Mass.

Chillón, M. *et al.* (1995) 'Mutations in the Cystic Fibrosis gene in patients with Congenital Absence of the Vas Deferens', *New England Journal of Medicine*, 332 (22), pp. 1475-1480.

Chmiel *et al.* (1999) 'Pitfall in the Use of Genotype Analysis as the Sole Diagnostic Criterion for Cystic Fibrosis', *Pediatrics*, 103(4), pp. 823-6.

Colin, A.A. *et al.* (1996) 'Pulmonary function and clinical observations in men with Congenital Bilateral Absence of the Vas Deferens', *Chest*, 110(2), pp. 440-445.

Cookson, W. (1999) 'Disease taxonomy – polygenic', *British Medical Bulletin*, 55(2), pp. 558-565.

Cudworth, A.G. (1976) The aetiology of diabetes mellitus, *British Journal of Hospital Medicine*, 16, pp. 207-216.

Cudworth, A.G. (1978) Type I Diabetes Mellitus, *Diabetologia*, 14, pp. 281-291.

Culard *et al.* (1994) 'Analysis of the whole CFTR coding regions and splice junctions in azoospermic men with congenital bilateral absence of epididymis or vas deferens' *Human Genetics*, 93, pp. 467-470.

Department of Health (2001) Press release: Boost for baby health, Monday April 30th.

Dumur *et al.* (1990) 'Abnormal distribution of CF ΔF508 allele in azoospermic men with congenital asplasia of epididymis and vas deferens', *Lancet*, 336, p. 512.

Friday, R.P., Trucco, M. and Pietropaolo, M. (1999) 'Genetics of Type 1 diabetes mellitus' *Diabetes Nutrition and Metabolism*, 12 (1), pp. 3-26.

Gale, E.A.M. (2001) The Discovery of Type 1 Diabetes, *Diabetes*, 50, pp. 217-226.

Genuth, S. (1982) 'Classification and Diagnosis of Diabetes Mellitus', *Medical Clinics of North America*, 66(6), pp. 1191-1207.

Gervais *et al.* (1993) 'Sweat chloride and ΔF508 mutation in chronic bronchitis or bronchiectasis', *Lancet*, 342, p. 997.

Goldspiel, B.R., Green, L. and Calis, K.A. (1993) 'Human Gene Therapy', *Clinical Pharmacy*, 12(7), pp. 488-505.

Goyder, E. and Irwing, L. (1998) 'Screening for Diabetes: what are we really doing?', *British Medical Journal*, 317, pp. 1644-1646.

Hedgecoe, A. (2002), 'Reinventing Diabetes: Classification, Division and the Geneticization of Disease', *New Genetics and Society*, 21(1), pp.7-27.

Hesslow, G. (1984) 'What is a Genetic Disease? On the relative importance of causes' in L. Noredenfelt and B.I.B. Lindahl (eds) *Health, Disease, and Causal Explanations in Medicine*, Boston: Reidal, pp. 183-193.

Hull, R.T. (1978) 'On getting 'Genetics' out of 'Genetic disease' in J. Davis, B. Hoffmaster and S. Shorten (eds) *Contemporary issues in Biomedical Ethics*, Humna Press, Clifton NJ, pp. 71-87.

Iannuzi, M.C. and Collins, F.S. (1990) 'Reverse genetics and cystic fibrosis', *American Journal of Respiratory Cell Molecular Biology*, 2(4), pp. 309-316.

Irvine, W.J. (1977) Classification of Idiopathic Diabetes, *The Lancet*, 2, pp. 638-642.

Jones, S. (1993) *The language of the genes: biology, history and the evolutionary future*, Harper Collins, London.

Keen, H. (1985) Limitations and problems of diabetes classification from an epistemological point of view in M. Vranic, C.H. Hollenberg and C. Steiner (eds) *Comparison of Type I and Type II Diabetes: Similarities and Dissimilarities in Etiology, Pathogenesis and Complications*, Plenum Press, New York and London, pp. 31-46.

Keen, H. (1986) What's in a Name? IDDM/NIDDM, Type 1/Type 2, *Diabetic Medicine*, 3, pp. 11-12.

Kerr, A. (2000) (Re) Constructing Genetic Disease: the clinical continuum between cystic fibrosis and male infertility, *Social Studies of Science*, 30(6), pp. 847-94.

Lawrence, R.D. (1951) 'Types of Human Diabetes', *British Medical Journal*, February 24, pp. 373-375.

Lissens, W. and Liebaers, I. (1997) 'The genetics of male infertility in relation to cystic fibrosis', *Baillièrre's Clinical Obstetrics and Gynaecology*, 11(4), pp. 797-817.

Mandrup-Poulsen, T. (1998) 'Diabetes: Recent Advances', *British Medical Journal*, 316, pp. 1221-1225.

McKusick, V.A. (1969) 'Lumpers and splitters, or the nosology of genetic disease' *Birth Defects*, 5, pp. 23-32.

Mearns, M.B. (1993) 'Cystic Fibrosis: the first 50 years' in J. Dodge, D. Brock and J. Widdicombe (eds) *Cystic Fibrosis – Current Problems Vol. 1*, Wiley and Sons, London, pp. 217-250.

National Diabetes Data Group (1979) Classification and Diagnosis of Diabetes Mellitus and Other Categories of Glucose Intolerance, *Diabetes*, 28, pp. 1039-1057.

Office of Technology Assessment (OTA) (1998) *Mapping our Genes. Genome Projects: How Big, How Fast*, Johns Hopkins University Press, London and Baltimore.

Rosenstein, B.J. (2002) Cystic Fibrosis Diagnosis: New Dilemmas for an Old Disorder, *Pediatric Pulmonology*, 33, pp. 83-84.

Rosenstein, B.J. and Cutting, G.R. (1998) The diagnosis of cystic fibrosis: A consensus statement, *The Journal of Pediatrics*, 132(4), pp. 589-95.

Rosenstein, B.J. and Zeitlin, P.L. (1998) 'Cystic Fibrosis', *The Lancet*, 351, pp. 277-281.

Sawyer, S.M., Tully, M.M. and Colin, A.A. (2000) Reproductive and sexual health in males with cystic fibrosis: a case for health professional education and training, *Journal of Adolescent Health*, 28(1), pp. 36-40.

Seamark, C.J. and Hutchinson, M. (2000) 'Should asymptomatic haemachromatosis be treated?' *British Medical Journal*, 320, pp. 1314-1317.

Sharer, Swartz and Path (1999) Reply to letter, *New England Journal of Medicine*, 340(3), pp. 238-9.

Shepherd, M., Hattersley, A.T. and Sparkes, A.C. (2000) 'Predictive Genetic Testing in Diabetes: A Case Study of Multiple Perspectives', *Qualitative Health Research*, 10(2), pp. 242-259.

Smallwood, M. (1990) *Understanding Diabetes*, Houghton Miffins, Victoria.

Stern, R.C. (1997) 'The Diagnosis of Cystic Fibrosis', *New England Journal of Medicine*, 336(7), pp. 487-491.

Super, M. (1992) 'Milestones in cystic fibrosis', *British Medical Bulletin,* 48(4), pp. 717-737.

University of Florida (2002) 'Press release: American Diabetes association and the University of Florida announce plans for $10 million statewide infant screening initiative for type 1 diabetes', January 2002.

Wachbroit, R.S. (1994) 'Distinguishing genetic disease and genetic susceptibility', *American Journal of Medical Genetics*, 53, pp. 236-240.

Wareham, N.J. and O'Rahilly, S. (1998) 'The changing classification and diagnosis of diabetes', *British Medical Journal*, 317, pp. 359-360.

Winter, W.E., Nakamura, M. and House, D.V. (1999) 'Monogenic Diabetes Mellitus in Youth: the MODY Syndromes, *Endocrinology and Metabolism clinics of North America*, 28(4), pp. 765-785.

Zielenski, J. (2000) 'Genotype and Phenotype in Cystic Fibrosis', *Respiration*, 67, pp. 117-133.

PART V
CLONING AND XENOTRANSPLANTATION

13 What We Know and What We Don't About Cloning and Society

Sarah Franklin

One of the most interesting claims in Gina Kolata's *Clone: the Road to Dolly and the Path Ahead* comes at the very end of her highly informative and engaging narrative. According to Steen Willadsen, the embryologist whose work paved the way for Ian Wilmut's eventual success, humans may have already been cloned – accidentally. Now working part-time in a private IVF clinic in the United States, Willadsen suggests that a procedure for infertile men involving the removal of immature sperm cells that are then microinjected into the egg may result in the transfer of too much DNA. To mature, sperm cells must become haploid – they must shed one of their two sets of chromosomes. If they are immature when they are injected into the egg, and if the egg subsequently sheds its own nucleus, the resulting embryo will be a clone of the father. According to Kolata, Willadsen considers this occurrence unlikely, but not impossible. Indeed, he argues that it is simply a matter of time: "even the most unlikely event will eventually occur if you wait long enough".

What would it mean if this were true, if a human had already been cloned? Probably very little, which is probably why no one seems to have noticed this rather startling anecdote in what is still (April, 1999) the only single-authored book available about the cloning of Dolly the sheep. Maybe it wouldn't matter, because that would mean the first human clone was created unintentionally, and it appears a great deal of concern about cloning is not just about a new-found technological capacity, but the forms of human intention and choice that will shape its future. Or perhaps it is because there is no easy way to verify Willadsen's speculation, and so it will never make the headlines.

Two years after Dolly, the cloning debate offers a benchmark case-study of late-twentieth-century public debate about reproductive technology and the new genetics. In particular, this debate offers a chance to observe the relationship between highly technical biological science and the arguments of a wide range of critics, commentators and policy makers. Traditionally, the authoritative account of some new biotechnological possibility, such as gene therapy or IVF, begins with a recitation of 'the scientific facts'. Hence, for example, almost all official reports from advisory committees begin with pithy summaries of current scientific understandings, like preludes to the subsequent symphony of diverse viewpoints and opinions. Such scientific descriptions are presented as the neutral, objective, factual basis that constitutes a shared, undisputed territory.

However, one of the problems with 'cloning' is that the birth of Dolly challenged the stability of biological facts. Dolly jumped right out of the biology rule book, and this is one of the features of her birth that has caused anxiety. Her very existence is counterfactual. Or at least it used to be.

Only one published response to the cloning episode has responded to this paradox, and that is the highly informative report by the Wellcome Trust Medicine and Society programme, entitled *Public Perspectives on Human Cloning*, which presents the results of a specially commissioned study. Mutating a well-established method of qualitative research, the focus group, by convening serial sessions of the same group and introducing focus briefings, which provided a sophisticated account of the science of cloning, this study presents 'the scientific facts' in a more fluid, relational manner. "The aim is to measure attitudes and any changes during a longer deliberative process", note the study's authors, Suzanne King, Ian Muchamore and Tom Wilkie. In turn; the report offers the intriguing finding that public acceptance of cloning is not enhanced by public understanding of the new genetics. In fact, the more people understood about the science of cloning (which they proved very good at assimilating), the more distrustful they became: "as participants' awareness increased so did their concern and apprehension", the study concludes. Participants in the focus groups viewed regulation with scepticism, and "were unconvinced public opinion would have any effect on what research was done". While their expectations of medical research were "high", the study revealed frequent turns to "conspiracy theory" and found that "suggestions that secret research was taking place were common".

Two other British reports on cloning were also published in December 1998, from the joint Human Fertilization and Embryology Authority (HFEA) and Human Genetics Advisory Commission (HGAC) (1998) consultation exercise, and from the Farm Animal Welfare Council (1998). Both of these reports are more conventional, although the former offers a helpful summary of international law on cloning, and the latter provides a level of technical detail about the processes involved in animal cloning that is not found elsewhere. The HGAC/HFEA report presents the findings of a consultation exercise based on soliciting views from individuals and established organisations (200 responses were received) along with recommendations. The main proposal is a ban on cloning as a means of human reproduction, alongside a widening of licensing arrangements in order to explore the therapeutic benefits of nuclear transfer, including "the development of methods of therapy for mitochondrial disease" (this technique is both germline gene therapy and highly experimental). Similarly, the Farm Animal Welfare Council calls for a moratorium on the use of cloning by nuclear transfer in commercial agricultural practice, or its introduction as a routine technique, while calling for further research to improve this method's safety and efficiency.

Evident in all three British reports, and prominent in the 1997 Report of the US National Bioethics Advisory Commission on "Cloning Human Beings", are the problems posed by the term 'cloning' itself. As the HGAC/HFEA report puts it: "There are often difficulties over finding mutually acceptable and agreed definitions for even quite simple concepts". There are several reasons for these difficulties in relation to 'cloning'. To begin with, it has become an especially emotive term as a result of the image of clones and cloning from literature, such as Huxley's *Brave New World*. Gina Kolata helpfully documents more recent occasions in which cloning has "sullied science", such as the David Rorvik controversy in the 1970s. It is also technically confusing because 'cloning' is something of a basket category for a range of quite different phenomena. Some of these are as uncontroversial and ubiquitous as mitosis – the most common form of organic reproduction, which is asexual, and which not only occurs among micro-organisms and plants, but among vertebrates, such as fish, reptiles and amphibians. Other cloning techniques, such as the splitting of an embryo in half, are both naturally occurring and technologically induced. With the development of modern molecular biology techniques such as polymerase chain reaction, cloning has arguably become the single most important research technique in contemporary biogenetic science. Twins can be described as clones, if genetic identity is the measure of what a clone is, but cloning is also used to describe complex procedures, such as nuclear transfer, a form of cell fusion.

Cloning by nuclear transfer does not involve the production of a new offspring from a single cell, or bud, of an adult, the way a gardener creates a new hydrangea from a cutting. Several individuals are involved in this form of reproduction, including the nuclear donor, the egg donor and the surrogate, or sometimes two surrogates–one to culture a large number of fertilized oocytes *in vivo*, and another to carry two or three blastocysts to term. Some developmental anomalies are associated with the use of nuclear transfer in sheep and cows, and many unknown factors are suggested by the very few live births that result, and the presence of deformities affecting some offspring that do survive. Technically speaking, it is somewhat misleading to refer to the technique that produced Dolly as 'cloning'. As an eminent British biologist recently said to me, nuclear transfer is a very sophisticated piece of reproductive biology, but it is not cloning.

In the original article on Dolly in *Nature*, the terms 'clone' or 'cloning' do not appear. 'Clone' and 'cloning' were also not used by the research scientists Robert Briggs and Tom King for their famous experiments on frogs in the 1950s, in which they confirmed the viability of what they called "nuclear transplantation" (although John Gurdon did use the term 'clone' to describe his frog experiments in the 1960s). Arguably, the term 'cloning' belongs less to scientific than to popular discourse, where it has increasingly come to be used as a condensed signifier for the potential of genetic science to produce unnatural kinds. 'Cloning' may be a term scientists such as

Wilmut deliberately avoid, precisely because it has become a kind of popular shorthand for science 'gone too far'. Like the term 'test-tube babies, which is seen to carry a negative connotation, 'cloning' is associated with iconic images of Frankenstein monsters, Nazi medicine, and *The Boys From Brazil* – which are among the most commonly encountered idioms of public scepticism and distrust toward science, and especially the life sciences. This capacity for clones and cloning to function as potent symbols of the dangers of modern science, and indexes precisely the public mistrust of scientific research documented in the Wellcome report. As Gina Kolata writes: "Today, the public is both drawn to, and often frightened by, the new powers of medicine and technology". In contrast to the generally positive and hopeful view of science and scientists in the decades just after the Second World War, which saw the discovery of penicillin, the development of vaccines for smallpox and polio, and the dawn of the space age, Kolata argues, "this is an era that many have described as anti-science", in which the public increasingly "views scientists as isolated technocrats who rather high-handedly attempt to argue from positions of authority".

The visceral and volatile effects of the very term 'cloning' may be one source of the unexpected public reaction to the birth of Dolly the sheep. At the Roslin Institute, for example, public reaction to Dolly came as a total surprise. Especially in the wake of the dim fanfare surrounding the announcement of the births of the cloned sheep Megan and Morag a year earlier, no one in out-of-the-way Roslin anticipated even vaguely the scope of international response, its speed, or its intensity. Understanding the impact of cloning requires appreciating its symbolic capital, as well as its biological significance. Indeed, the cloning debate could not have provided a better indication of how powerfully 'biological facts' operate as natural symbols stabilizing the order of things, or destabilizing them. It requres a U-turn from the literal to the symbolic to explain how quickly a novel form of impregnating a sheet could generate such intimate anxieties about human kinship, personhood and identity.

Once Dolly had become international headline news, the stage was set for what has become not only a public debate, but a global one, with echoes of the 'miracle birth' of Louise Brown in 1978. Since Dolly was announced two years ago, the cloning of mice, monkeys, cows and even humans have been announced from laboratories in Europe, North America and Korea, and governments around the world have strained to keep pace with the rapid developments in the field of cloning. A book that is yet to be written could very productively compare these international responses at the level of media and legislation.

Three anthologies published in the USA in 1998 offer a selection of pieces by various commentators on cloning and public reactions to it, and give a good sense of the range of views expressed both in the media and before Congress. While many of the positions are familiar, there are some interesting juxtapositions. In *The Ethics of Cloning*, published by the American Enterprise Institute (Kass and Wilson, 1998), for example, two conservative academics offer opposing views of cloning – both based on the 'natural facts of family life'. For Leon R. Kass, cloning is wrong because it violates "the profundity of sex". As if writing in support of a fundamentalist Christian view of life as a divine gift, only with a lot more emphasis on gametes instead of God, Kass argues that the culmination of coitus in procreation is a deep and meaningful mystery, the sanctity of which we violate at the cost of our very humanity. Right back at Kass comes James Q. Wilson, arguing it is in fact not the heterosexual bond, but the mother-child bond that is "the most powerful in nature", and consequently cloning is fine, as long as it is restricted to within marriage. This text illustrates neatly the flexibility of biological determinism as symbolic currency, in perfect concert with the constant remoulding of biology itself to which it responds.

The bioethics community is represented in *The Human Cloning Debate* (McGee, 1998), edited by Glenn McGee, a member of the Penn Centre for Bioethics. Two opening chapters, by *Nature* journalist Potter Wickware and philosopher Ina Roy, offer introductions to the science and ethics of cloning, respectively. Wickware's busy tour of biology is full of vivid metaphors for DNA and cloning (he asks "is a reissue of David Copperfield using the original 1850 plates new or old?" and concludes with a demonstration of the difference between genetic essentialism prominent in many discussions of cloning). Ina Roy stresses a fundamental difference between consequentialist approaches to ethics, which stress outcomes, such as utilitarisnism, and 'deontological' frameworks, which stress intention, as in the work of Immanuel Kant, who stressed the morality of ends in themselves. John Robertson, a legal scholar, combines these two approaches in his argument that cloning should be allowed as an expression of procreative liberty for couples, as long as they intend to rear the child themselves, thus limiting the practice to those with good intentions, while preserving the maximum benefit for the widest range of people, including the child. Arthur Caplan considers the question of whether ethical objections have ever stymied scientific and technological progress, arguing that, while it is difficult to identify a case where they have, this is perhaps to underestimate the slow effect of ethics in shaping public policy, which is he suggests "more akin to detecting the processes of evolutionary change, being aware of barometric pressure, or being alert to the presence of gravity". If there were ever a better case for cloning for ethics to catch up with public policy, he suggests, it would be hard to find. A chapter by Glenn McGee and Ian Wilmut argues for an adoption model to prepare the way for cloning as a means of creating new families. Philip Kitcher also supports cloning-for-families on behalf of stable lesbian couples who would like to have a child, and who could, if one partner donates the egg and the other the nucleus, more closely emulate the heterosexual ideal of conjugal and procreative unity (arguably not the most widely shared aspiration among lesbian couples). This volume also reprints the Leon Kass essay, "the Wisdom of Repugnance", as well as several short anti-cloning pieces representing religious traditions including Islamic, Judeo-Christian, and Buddhist perspectives, and the Pro-Life movement. A critique of religious dogmatism by bioethicist Ronald Lindsay concludes this section of the book, which itself concludes with a short fictional piece by Richard Kadrey. Despite many typographical errors and a hurried unevenness of tone and clarity, this is a useful anthology that contains many pieces written for submission to Congressional hearings and consultation exercises, as well as helpful extracts reprinted from the Report of the National Bioethics Advisory Committee (NBAC).

Clones and Clones: Facts and Fantasies about Human Cloning (Nussbaum and Sunstein, 1998) is also a US-based publication, edited by Martha Nussbaum and Cass Sunstein. In this volume, too, are reprinted extracts from the NBAC report, which has been widely praised for its clarity and scope. Helpfully, this volume also includes the original *Nature* article, "Viable offspring derived from fetal and adult mammalian cells". Twenty additional chapters cover a wider range than the McGee anthology, including contributions from sociology, lesbian and gay studies, feminist theory, and history, as well as the staple triumverate of theology, philosophy and law. In stark contrast to the telegraphic style of Wilmutt and his colleagues, biologist Stephen Jay Gould provides a characteristically lucid, informative and engaging chapter in which, like Lewontin and Wickerware, he reminds us of the many differences between genetics and biology. Protesting that the nature-nurture dichotomy is as unhelpful as it is tenacious, he compares the pendulum-like swings of public opinion from one extreme to another to the vagaries of fashion, and argues we are currently witnessing a resurgence of genetic essentialism. Unfortunately, few of the chapters in this volume are as well-written or thoughtful as Gould's. Richard Dawkins writes about a bad experience he had on a radio talk show, which confirmed his views that many religious spokespeople are lacking in intelligence. Andrea Dworkin imagines cloning her cat but decides it would be complicit with "the absolute power men have wanted over reproduction and have destroyed generations upon generations of women to approximate". William Eskridge and Edward Stein defend cloning as "the next logical step in queer people's formations of families of choice", and thus a more "queer friendly society" in general. Incredibly, Eric and Richard Posner provide a numerical tabulation of "the private benefits and social costs of human cloning", illustrated by a table entitled "Genetic endowment of offspring under alternative reproductive regimes" (which demonstrates, in economic terms, "the payoffs" of different kinds of mating).

Wendy Doniger's historical account of proto-clones in mythologies and traditional beliefs about conception is indicative of how well the cloning question is served by widening the frame historically and culturally. William Miller's chapter on sheep jokes, cloning and the uncanny is also refreshingly original, and suggests that another worthy project would involve collecting all of the sheep cartoons from the past two years. On the whole, however, beside too much earnest philosophical explanation is too little attention to cloning from further afield, from other disciplines, and in particular from a cultural perspective. Popular film, for example, is full of images of clones, as is literature and science fiction, and these sources can be very productively explored (although it should be noted that fictional pieces are included in both anthologies). Contemporary art is another source of powerful imagery of changing definitions of life and death, such as the work of Damien Hirst or Helen Chadwick, in which sheep and human embryos are respectively deployed in pieces directly addressed to science. Other gaps are theoretical. It is odd, for example, not to see at least some reference to the work of Foucault, one of the century's most prominent interpreters, on biopower and the changing interrelation of life, labour and language evident in the new genetics. Curiously, the fact that the cloning of Dolly was financed by a pharmaceutical corporation also goes largely unexamined, as does the creation of her transgenic companion, Polly, and the unusual forms of property potection through which they are both owned as bio-wealth.

Without undervaluing some truly thoughtful work and some excellent science writing, the cloning debate as it appears in these volumes contains many arguments we have seen so often before they have become clichés. This is especially ironic, given the way in which cloning has itself been decried as a source of endless similarity, the reduction of individuality and the antithesis of creativity. Repeated again and again, concepts such as procreative liberty, the sanctity of the nuclear family, the right to a unique genetic identity, the value of scientific progress, the need for rational argument, and the maximization of personal freedom and happiness end up sounding more like stereotypes than analysis. At the core of the cloning debate is a familiar question about limits: everyone agrees we should have them, but it is very hard to agree on how they will be defined, by whom, and on what terms. Some people, like Leon Kass, seem to think this is almost a matter of instinct. Others, such as Dawkins, want only pure reason as a guide. Historians look to the past, philosophers imagine the future, lawyers rehearse courtroom dramas. One begins to feel it is time to start being a bit more imaginative.

The appealing feature of the Wellcome report is that it begins and ends in uncertainty: there is something very appropriate and reassuring about examining how much we do not know about public responses to Dolly. One of the main suggestions of the Wellcome report (which, like all qualitative research, is not representative, but may be indicative) is that members of the British public believe they are largely ignorant about what is going on in science laboratories. They are distrustful, and do not believe what they are told by the government, by policy makers, or by regulatory bodies. They feel scientists are themselves often high-handed and dismissive of their concerns, and they feel their own opinions matter very little in debates over controversial subjects such as cloning. Why is this message important to the debate about cloning, or scientific innovation more generally? Because it suggests that the problem is not that a new kind of cloning has become possible, but that the social process of coming to terms with such developments is compromised by the reproduction of persistent forms of inequality and exclusion. This view is also supported by the Farm Animal Welfare Council, who recommend "as a matter of some urgency" that, in addition to promoting public understanding of issues such as cloning, it is also necessary "to improve politicians' and scientists' understanding of the fundamental public concerns which undoubtedly exist".

The inequality between scientific expertise and 'lay' opinion is re-enacted in the generic format of official reports, which privilege authoritative versions of scientific facts before turning to the controversial matter of how to govern them. This format is itself a kind of representational technology, not only securing a prominent boundary between science and its publics, but sending a message about the power relations that structure their relation. Most accounts of cloning define it in relation to the emergence of a new form of scientific and technological capacity.

But 'the cloning debate' is about society, not technological systems. Many people who have no scientific training whatsoever may well know a great deal about what society is, how they think it should be, and how people should treat one another. Many people with no scientific training have children, we all have parents, it matters to most of us how animals are treated, we know not everyone thinks the same things we do, we agree there have to be certain rules about how things are done, even if they sometimes are broken, or need to be changed.

The Wellcome report thus identifies an important and persistent social deficiency, and it is to be commended both for documenting the degree of public mistrust of genetic science, and for developing means of exploring it in greater depth. Scientists who already feel beleagured, misunderstood and underfunded may not embrace this finding, but it is significant the study is itself scientific: it is empirical, analytical and rigorous. In a sense, the Wellcome Trust conducted an experiment, and they have certainly identified some significant findings. An implication of their findings is that the 'problem' posed by cloning is often misplaced: the real challenge is not what to do about it, how to regulate it, whether to permit it or not, but how more democratically to conduct a less exclusive conversation about such questions. And this is clearly something very difficult to do, almost unimaginable, like getting rid of prisons (even though everyone knows they don't work). But then, the first step to cloning was imagining something very clever that no one else thought would ever succeed.

Amidst the current debate in Britain about genetically modified foods, and the continuing BSE controversy, it is clear that agricultural technologies are becoming a subject of increasing public concern, and this point is especially clear in relation to the cloning debate, which concerns a technique developed for sheep breeding. In contemporary British society, as elsewhere, the politics of food production, animal breeding and human health have become profoundly inter-woven along some novel connecting threads, such as Monsanto's relationship to chocolate-covered biscuits, Prince Charles's affection for beef-on-the-bone, and a frozen cell line's connection to a young Scottish ewe. These are the same newly meaningful connections that lie at the heart of a new politics of the supermarket aimed at goals such as preserving old English orchards, through mechanisms such as the Soil Association's kite-mark for organic foods. At the root of such quotidian consumer activities as selecting organic foods (now a world-wide consumer trend) is a sensation of permeability and exposure to risk – to hidden risks that are only further obscured by a noisy gaggle of befuddled experts. It is a mistake to discount such concerns as manifestations of ignorance or misunderstanding. It is equally a mistake to underestimate the importance of an emergent symbolic vocabulary in which cattle feed stands for brain diseases or soy beans represent corporate conspiracies. The way in which biological science has increasingly become part of a sophisticated public language of protest and critique directly challenges the view that expert knowledge can be siphoned out of the fray and set alongside it, or above it, as though shepherd to an unruly flock. That separation is as out-of-date as the idea that you can't combine an adult cell nucleus with an egg to produce a higher vertebrate. The fusion has happened, and like Dolly, the newly biologically literate British public is stamping its feet when it is annoyed with its dinner.

References

Farm Animal Welfare Council (1998) *Report on the Implications of Cloning for the Welfare of Farmed Livestock*. Surrey, UK: Ministry of Agriculture, Fisheries and Food.

Human Genetics Advisory Commission and the Human Fertilization and Embryology Authority (1998) *Cloning Issues in Reproduction, Science and Medicine*. London: Department of Trade and Industry.

Kass, L.R. and Wilson, J.Q. (1998) *The Ethics of Human Cloning*. Washington, DC: American Enterprise Institute.

Kolata, G. (1997) *Clone: the Road to Dolly and the Path Ahead*. London: Allen Lane/The Penguin Press.

McGee, G. (ed.) (1998) *The Human Cloning Debate*. Berkeley, CA: Berkeley Hills Books.

Nussbaum, M.C. and Sunstein, C.R. (eds) (1998) *Clones and Clones: Facts and Fantasies About Human Cloning*. New York: W.W. Norton.

Wellcome Trust, Medicine and Society Programme (1998) *Public Perspectives on Human Cloning*. London: The Wellcome Trust.

14 Containing Contradictions: Debating Nature, Controversy and Biotechnology

Nik Brown

Introduction

Current social theoretical commentary identifies ambivalence and contradiction as a normative feature of contemporary culture. That is, far from being extraordinary or even avoidable, contradictions pervade individual and institutional experience in ways which are increasingly difficult to overlook (Bauman, 1991). For some, this is a logical consequence of the indeterminacies of risk culture (Beck, 1992), where action proceeds on the basis of knowledge which is permanently open to revision. Contradictory knowledges seem to vacillate so quickly that they sometimes take hold simultaneously rather than consecutively (Virillio, 2000). For others, the very boundaries (between say science and culture, nature and politics) that allowed modernity's contradictions to operate now appear untenable (Latour, 1993). A state of contradiction might be said to be a situation in which subjects '...encounter alternatives, such that the choice of one alternative necessarily entails some loss of, or places some limit on, the use of the other. Contradictions are thus contraint-generating, inhibitory situations...' (Gouldner, 1980, p.168). More importantly, contradictions 'imply impending change precisely because they proliferate pathologies of action and communication, multiply costs, and hence diminish the advantages of pursuing any course of action' (ibid).

In addition to being an inescapable feature of most people's lived experience, contradiction is itself a site of important analytical enquiry. Under what conditions are contradictions sustainable and by whom? What forms of action or inaction are inspired by the ambivalent awareness of contradiction? The purpose of this chapter is to understand what happens when disputants hold and promote quite contradictory arguments about nature and morality, specifically within the context of debates about biotechnology. At the centre of this account are two familiar, though complex, figures. The biotechnology industry on the one hand and opposition NGOs on the other. A number of Science and Technology Studies (STS) scholars have focused on the relationships between formal scientific institutions and various opposition constituencies. Invariably, the former are presented as disproportionately privileged in being able to generate social moralities which are endorsed by laboratory artifice and corresponding accounts of nature. On the other hand, pressure group organisations are often depicted as less able to represent themselves as producers of scientific knowledge, particularly in the absence of a laboratory research capacity. When such groups successfully destabilise scientific enterprises, it is often on the unequal basis of social questions about the foundations of trust or by employing (often quite marginal) counter-scientific claims.

The management of contradiction, and the consequences that follow from holding simultaneously conflicting positions, differ significantly for proponents and opponents within contemporary debates on biotechnology, particularly in respect to contradictions in their moral and scientific repertoires. Pressure group organisations for instance enjoy enormous flexibility in being able to successfully deploy fundamentally contradictory scientific and moral arguments. On the other hand, such contradictions can be catastrophic for scientific institutions, this analysis suggests, because they are more usually judged in relation to idealised and inflexible symbolic markers of authoritative expertise.

Whilst this argument might have some relevance to wider debates in biotechnology, it is situated within a specific case context. The transplantation of organs, tissues and cells across species boundaries is currently an intensely disputed topic in the regulation of contemporary transgenics. Debates have crystallised around a number of issues including the indeterminate future therapeutic efficacy of the approach and acute concern about the threats of transpecies disease. Also, xenotransplantation raises questions about the management of human and nonhuman identity and the ever-changing basis of the relationships between humans and other animals. Even more problematic is the difficulty of determining whose voice should be trusted in volatile disputes where intense convictions about morality, science and risk intersect with strong commercial pressures. The complexities of the case have fascinated many keen observers of biotechnology in the public sphere. It touches on questions about the adequacy of regulatory provision to take account of the views of 'situated public' (Welsh and Evans, 1999); the anthropological significance of disgust discourses (Brown, 1999) and approaches to the problem from moral-philosophical perspectives (Hilhorst, 1998).

At the centre of these contestations are a number of constituencies including industry, regulators, and various advocacy clusters including patient groups and animal welfare (and rights) organisations. This chapter addresses the way in which these debates are varyingly negotiated by two of these key constituencies. First, a number of biotechnology and pharmaceutical firms have invested heavily in the field, particularly in the mid 1990s though somewhat more modestly of late. On the other hand, these developments have been vigorously opposed by a number of animal advocacy NGOs. Both groups have drawn upon a wide range of moral and scientific resources with which to contest or defend the immediate and long-term future of xenotransplantation. This analysis maps the uses made of both morality and science by these constituencies and identifies striking areas of contradiction in their arguments. The analysis builds on previous discussions whereby moral and scientific discourses have been found to operate as the basis for structuring relations of difference and similarity between humans and other species (Brown, 1999). The case is also used to comment on science studies accounts of conflicts between scientific and pressure group organisations.

Science, Morality and Controversy Studies

Before proceeding, it is necessary to refine what is meant by morality and science for the purposes of this paper. Drawing on the work of Gieryn, science and morality emerge iteratively through the 'boundary work' in which they are constituted (Gieryn, 1983). Such divisions routinely assign selected characteristics to 'science (that is, to its practitioners, methods, stock of knowledge, values and work organisation) for the purposes of constructing a social boundary that distinguishes some activities as 'non-science' (1983, p.782).

The shape of controversies between different groups rests on how various attributions are stabilised, often around putative distinctions between nature, science and society-morality. For illustrative purposes, the demarcation of 'ideal' science might rest on the disinterested witness of scientists to nature using invariant objectivism as the basis for reliable claims about a world 'out there' separate from the cultural, moral and commercial one occupied by humans 'in society' (Shapin and Schaffer, 1985; Latour, 1993; Haraway, 1997). By inference, an equally idealised morality might refer to a subjective social sphere in which relatively impermanent values are formulated sometimes by consensual or democratic means negotiated between heterogeneous parties. What counts as the basis of either morality or science will always vary from one debate to another. Consequently, making sense of a controversy should depend more on how those categories are mobilised in practice by disputants thus avoiding the analytical use of pre-given scientific or moral typologies.

In practice, the congitive authority of science, claims Gieryn, stems from the way its claimants construct a space to appeal both to nature and morality without mixing-up either of the categories (1983). This apparent contradiction or 'Janus face' (Latour, 1987; 1993) is the basis of science's privileged authority. If the boundary work is successful, troublesome actors are denied access to inclusion in the boundary of science and relegated to using social resources like morality alone. On the other hand, the privileged category of science entails certain constraints particularly with respect to the way contradictions within the rhetorical space of science are managed. As the paper will go on to illustrate, claims to scientific identify entail certain normative conditions, like the requirement to maintain the appearance of consistency, thus reducing the capacity of claimants to accommodate contradictions when they arise.

Several other approaches to the dynamics of conflicts address the way 'non-scientific' constituencies, such as pressure groups, utilise various moral and natural science resources. In Nelkin's accounts of controversy, it is the underlying values, politics and morality of developments in science which serve as the principle negotiating territory within which such debates are fought out (Nelkin, 1984; 1992; 1995). Indeed, whether it be the teaching of evolutionary theory, vivisection, or the status of the foetus, 'the most intense and intractable disputes concern the social, moral, or religious implications of a scientific theory or research practice' (1995, p.447). When science does enter into the analysis of a controversy, it is often in terms of NGOs strategically drawing upon scientific expertise in order to deploy a dissident rendering of the technical terms of a debate thus contributing to 'public confusion over disputes between scientists' (ibid.). As Nelkin argues, such conflicts tend to raise public questions about the different social motivations responsible for producing competing scientific testimony. Accordingly, in drawing attention to conflicts amongst 'experts', NGOs are able to weaken their opponent's claims to objective expert evidence. But, as producers of scientific evidence in their own right, NGOs are often at a disadvantage in having an ambiguous scientific status and having to draw upon secondary, and often marginal, scientific knowledge.

Similarly, Wynne has explored the way scientific claims come to be disputed by NGOs and other publics on 'social questions about the basis of trust' and inadequate appreciations of local contexts of use (1995, p.377). Disputes over the nuclear contamination of Cumbrian sheep herds following Chernobyl offers insights into the dynamics between scientific institutions and public constituencies (Wynne, 1989; 1991). Scientists from the Ministry of Agriculture, Fisheries and Food (MAFF) based their estimates of how quickly the polluting caesium would fall to normal levels on soil assessments of lowland pastures. Farmers however, reserving lowland areas for winter grazing, would usually graze their sheep on upland pastures where clay soils inhibit caesium depletion. In this case, farmers' local expertise in land management and the habits of the sheep, not the use of counter scientific claims, discredited trust in MAFF's recommendations and their handling of the crisis.

Yearley comments more directly on the uptake of science by environmental NGOs or 'the use of science as ecological critique' (1995, p.459; 1992). Whilst such organisations are often undermined by an inability to have ecologically focused science accepted as scientific orthodoxy, they are often stronger in being able to point to conflicts within and between 'established' scientific communities. In common with Beck's thesis whereby science is deconstructed by being both source *and* solution to contemporary environmental hazard, the paradox also undermines the objective integrity of scientific orthodoxy and helps to level the epistemological terrain in which ecological science has to operate (Beck, 1992). The motivation behind NGO's use of scientific discourse is often based on the ineffectual status of environmental moral reasoning: 'the bodies with which environmentalists must deal respond better to technical and apparently objective claims than to arguments couched in spiritual or moral discourse' (Yearley, 1995, p.464). There are, however, drawbacks for environmental NGOs using science. For example, science often fails to deliver the evidence or 'decisiveness and moral certainty that activists desire' (ibid. p.461). Also, just as the revocability of scientific findings can undermine trust in adjudicating scientific institutions, ecological science too can often fail to produce findings which are durable enough to further an NGO's moral objectives. The use of science can also alienate constituencies within the ecology movement by contested symbols identified as the source of the problem.

In each of these cases, public opposition constituencies are able to use a range of resources with which to undermine developments in science and technology. These resources are largely oriented towards social, moral or value questions about certain developments. On occasions, such groups are also able to strategically borrow upon and deploy conflicts between experts on the technical or scientific terms of a case. In turn, this tends to highlight political or commercial motivations which have been instrumental in shaping opposed (largely dominant) scientific viewpoints, thus weakening claims to disinterested neutrality, consistency and objectivism.

As I will point out, many of these dynamics can be clearly seen in the way competing versions of the moral and social world are used to organise positions in the xenotransplantation debate. However, the case also highlights the flexibility which enables animal advocacy groups the ability to strategically deploy moral and social arguments which are fundamentally contradictory. Such contradictions are much more difficult for scientific institutions to maintain given the material and symbolic resources from which they derive their expertise in the first place. As such, the paper suggests that, whilst many environmental and animal advocacy NGOs have sought to present themselves through the prisms of scientific expertise, given the inflexibility of the discursive space of 'science', it is not always in their interests to do so.

Xenotransplantation: firms and firm opponents

By way of background, the arrival of mammalian transgenic techniques in the mid to late 1980s, coupled with nuclear transfer cloning much more recently, inspired large scale commercial investment in xenotransplantation by several international pharma companies. The overwhelming focus of research and development has been directed at overcoming immuno-rejection processes between unrelated species. However, virologists' concerns in the mid 1990s, that the approach might become a vehicle for potentially catastrophic transpecies disease incidents, has been responsible for subsequently far tighter regulation and the consolidation of international governance structures. Needless to say, the innovation history of xenotransplantation has been played out against numerous recent risk crises, characterised by ferocious antagonism between industry and animal advocacy NGOs. What follows is a brief characterisation of these two constituencies.

Industrial activity: Amongst the first wave of biotech ventures in the mid 1980s, Imutran Ltd specialised in the production of transgenic pigs and became a wholly owned subsidiary of Novartis in the mid 1990s. Novartis ceased European XTP-related operations in 2000, relocating a portion of its research to Boston MA under the new operating name Immerge. Immerge joins a number of US based companies which continue to be active in the field including Nextran, Genzyme Corp, Alexion Pharmaceuticals and PPL Therapeutics.

Animal Advocacy: For a number of British NGOs, XTP became a key focus of campaign attention in the mid 1990s and prior to the closure of Imutran Ltd. Xenotransplantation Concern ('XtC') represented an alliance of NGOs, served as a means of bringing together the campaign capacity of groups such as Uncaged Campaigns, Animal Aid, The British Union for the Abolition of Vivisection (BUVA), The National Anti-Vivisection Society, Compassion in World Farming (CIWF) and The Animal Welfare Foundation.

Contradiction and Science: clinical efficacy and transpecies disease

Let us explore some of the scientific terms of this debate first. The future of XTP depends upon proponents successfully convincing regulators, the public and the clinical community that the twin risks of rejection and disease have been sufficiently reduced to permit clinical application. Industry research has, in the first place, been overwhelmingly directed at producing some degree of immunological similarity between their transgenic pigs and potential human hosts. That is, the promise of clinical efficacy depends upon the capacity of the xenograft to be 'recognised' by the host's immune system.

In briefly reviewing the technical terms of xenografting, molecules on the surface of the graft (antigens) indicate whether the matter belongs to the body or is different to it. If recognised as 'foreign', antibodies adhere to the graft's antigens. In turn, this can trigger a 'complement cascade', a chain reaction in which a number of blood proteins (complement) puncture the cells of a transplanted graft. The approach commonly taken by the XTP biotech companies has been to develop transgenic pigs in which porcine complement is substituted by human complement (DAF, CD59 or MCP) thus, the argument goes, preventing a hyperacute rejection response (Cozzi and White, 1995; McCurry, 1995; Thompson, 1995). The UK Company referred to earlier, Imutran Ltd, produced pigs in the mid 1990s which express human DAF complement proteins. More recently, research has turned to the role of a carbohydrate on the surface of pig cells called alpha 1,3-galactose (alpha gal). Whilst humans lack alpha-gal, we possess antibodies against it because of prior exposure to bacteria bearing the same sugar. In early 2002, a number of teams (including PPL and Immerge) announced the production of animals from which the alpha gal gene had been deleted (*Nature*, 415, 103-104). Also, the production of similarity between donor and host species depends on the use of immunosuppressive agents like cyclosporin, produced by the parent company of the former XTP Company Imutran. Indeed, the use of transgenic porcine organs relies on an extension of the technologies of immunosuppressive similarity to deal with other functions within the immune system. In addition to the T cell antibody responses stimulated by transplanted human organs, transgenic porcine tissues also trigger antibodies made by B cells. Suppressing two major aspects of the immune system requires the development and use of a new class of drugs called macrolides.

These then are the technologies for the production of similarity and the pacification of physiological difference within the rubrics of expert laboratory and experimental practice.

> PPL Therapeutics Plc ... is pleased to announce it had produced 'knock-out' piglets which were born as a result of using nuclear transfer (cloning) and PPL's patented gene targeting technology. A 'knock-out' pig has the specific gene that leads to the human immune system rejecting pig organs inactivated.
>
> (PPL Press Release, 2 January 2002).

> In October last year, Imutran presented new research on the development of transgenic pig organs, showing that hyperacute rejection had been overcome – the major hurdle in the development of animal organs for transplantation into humans.
>
> (*Imutran* press release, 10 December 1996)[1]

> "Rather than taking the pig and making sausages", says Paul Herring, Sandoz's head of pharmaceutical research, "you could take the cornea, kidney and heart. After all, many pig organs are remarkably similar in structure to human organs."
>
> (*Finance Weekly*, 18.7.95)

Turning to the animal advocacy NGOs, their arguments have traditionally disputed experimental findings extrapolated from research animals to humans and convincing regulators of their dissimilarity on scientific grounds (Elston, 1994; Jasper and Nelkin,

1992; Michael and Birke, 1995). In the context of XTP, the argument is adapted to contest the claims that transgenic animal organs will function as viable long-term substitutes for diseased human organs and tissues. Here, much of the campaign literature refers to high profile cases where previous xenogeneic clinical applications have failed as a consequence of organ rejection. A particularly salient example of this are the frequent references to the 1984 operation in which a fourteen-day-old infant was xenografted with the heart of a non-transgenic baboon. The story of 'Baby Fae', as she came to be known, became a harrowing media spectacle in the weeks before and after the child's death twenty days following surgery. Since then, the incident has persistently figured in present day debates surrounding the contested clinical efficacy of XTP and has been borrowed into the campaign literature of animal advocacy NGOs. In all, the argument asserts the immunological dissimilarity of humans and nonhumans drawing upon previous clinical research cases and secondary scientific findings:

> ...all transplanted organs are liable to rejection – the body's own defence mechanism's attempt to destroy 'foreign' organs. ...[XTP] brings far greater dangers such as hyperacute rejection, an extremely strong reaction which destroys the new organ within minutes. ...introducing human genes into pigs may avert this acute response, but the pig organ will still face other rejection processes which scientists believe will be stronger than those faced by human organs. Finally, even if the organs survived they simply may not work successfully within the human body.
>
> (*Uncaged Campaigns*, 22 January 1999)

> There are other problems with xenotransplants. For example, will the donor organs age at the same rate as the rest of the human body?
>
> (*Animal Aid*)[2]

> While a pig organ may appear, superficially, to resemble a human organ, once we examine more closely the role that organs play in the workings of the whole body, then subtle dissimilarities will take on major significance.
>
> (*Uncaged Campaigns*, Xenotransplantation Report)[3]

On the other hand, animal advocacy NGOs have also begun to exploit scientific debates about the dangers of pathogenic risk. In reducing the risks of rejection by increasing the similarities between donor and host species, the biotech companies must now contend with the dangers of too much similarity. This is the contradictory double bind by which shared immunological pathogens including porcine endogenous retroviruses (PERVs), bacteria, fungi, prion proteins, etc. Any number of such entities, called 'zoonoses', have entered late twentieth century parlance as endangering expressions of human-nonhuman similarity and are borrowed into the animal advocacy scientific arguments: BSE and CJD; new evidence surrounding the etiological origins of the HIV virus; recent outbreaks of chicken flu in Hong Kong; etc:

> [XTPs] carry a potentially devastating risk in that a currently unknown animal virus could trigger a new plague when it passes to human beings... The outbreak of flu, which killed more than 20 million people in 1918/19, probably spread to humans from pigs. And recent test tube studies show that pig viruses will replicate in human cells. So this is not a theoretical problem.
>
> (*Animal Aid*)

> The genetic engineering of the pigs to prevent rejection, ironically, could make it easier for viruses to infect a recipient of pig tissue and then infect contacts of the patient. ... When we add to the equation the effects of large doses of immunosuppressants, likely to be bigger than currently administered to allograft [human-to-human] recipients, then the probability of viruses transferring from pigs to humans increases.
>
> (*Uncaged Campaigns*, 22 January 1999)

Promoters of XTP have responded by commissioning research with which to diffuse the disease threat. One of the most recent studies, conducted by Imutran's parent company Novartis, examined blood samples of 160 patients previously exposed to living pig tissues through a range of then permitted clinical applications (Paradis *et al.* 1999).[4] It found that none of the patients had been infected with porcine endogenous retroviruses (PERVs) including patients who had been immunosuppressed. Although 23 patients had porcine DNA circulating in their blood (a phenomenon called 'microchimerism') and four patients showed a PERV-stimulated antibody response.

Not surprisingly, interpretations of the study's findings have varied considerably. One virologist questioned the adequacy and sensitivity of the antibody-based assays used to detect viruses and whether the study ignored possible viruses harboured elsewhere in the human body (Hopkins, 1999). Imutran itself concluded more positively that the findings 'support the use of closely monitored clinical trials as an approach to assessing the safety and efficacy of using porcine cells, tissues, or organs therapeutically in humans' (Paradis, *et al.* 1999, p.1240). Retrospective studies like this illustrate the powerful capacity of clinical research networks to gain access to and reinterpret former experimental knowledge in pressing the case for future events which might otherwise be ruled out. Similar pressures have led to intense policy-making activity to govern the secondary re-use of tissue samples by commercial ventures for purposes other than those originally given by the patient. It might be contended that this illustrates a retrospective governance orientation, or rather a case of the new biologies hurtling into the future with its eyes fixed firmly in the rear view mirror.

The importance of Novartis' line of reasoning for our discussion here is the way it shifts the relations of similarity and difference between humans and other species by suggesting a division in the disciplinary basis of the scientific debate. The 'boundary work' distinguishes between two disciplines, virology and immunology, and looks a little like this: promoters continue to advance claims to *immunological* similarity (the argument for therapeutic efficacy) whilst mobilising *virological* evidence of dissimilarity (the argument against pathogenic risk). Attempts at this kind of boundary work are necessary if contradictions in the similarity of porcine donors and human hosts are to be managed. The animal advocacy NGOs' response is to maintain pressure on the logical contradiction of humans and pigs sharing the same embodiment without risk. That is, they seek to unpick the boundary work between virology and immunology:

> The organ is expected to reside in intimate contact with the patient's tissues and blood for months if not years at a time...
>
> (spokesperson for *The Hadwen Trust for Humane Research*)[5]

To draw these points together, both promotional and oppositional constituencies can be seen to generate relations of physiological similarity and difference which are contradictory. XTP biotech firms, in arguing for human-nonhuman similarity have exposed their technology to the dangers of too much similarity. Animal advocacy groups have contested claims to human-nonhuman similarity (in debates about animal models and the clinical viability of xenografts) but then rely upon claims of too much similarity when exploiting evidence of transpecies pathogens.

The way in which social interests intersect with, or even inform, these opposed versions of naturalised debate become more apparent when we turn to the social and ethical strategems of both groups.

Contradictions and culture: morality and species identity

Clearly, there is no consensus on the cultural in which to situate what counts as the proper province of morality. What concerns us here is the way 'morality' and 'culture' emerge, in the xenotransplantation debate, as a basis for organising relations of difference and similarity between humans and other species. So then, another boundary has come into play producing morality as a property of peoples' competing values as distinct from universal nature inscribed in scientific knowledge. Moral debate on technologies like XTP broadly includes different ways of reaching ethical decisions through prescriptive ethics (what logically follows should be done), consensual ethics (what we can agree needs to be done), or majority ethics (what most of us think should be done), etc. The mobilisation of personal or intuitive responses, feelings and emotions also come into play (Hilhorst, 1998). The point is to observe what the practical value of a distinction between the 'moral' and the 'natural' serves for actors, how the distinction is used and to what effect.

In cultural discourse, and in common with wider legitimations of the use of animals in scientific research, representatives of the biotech firms depend upon assertions of human-nonhuman moral difference. Here, the division between science and culture serves as a mutually endorsing, though divided ontology, whereby it is possible to assert human-nonhuman similarity in scientific debate and difference in moral debate. Pigs are used as a source of food and heart valves and this is taken to be adequate prior precedence for their being used as a source of donor tissues and organs. For example:

> How can you criticise the use of pig tissue for therapeutic procedures that save lives while at the same time accepting the existence of a ham sandwich?
>
> (*Imutran* director)[6]

In this case then, the relevant criterion for determining membership of the moral community is a hierarchical species arrangement. The rationale implied in the statement above, also mobilises certain ideal audiences, in this case a homogenised public for whom the eating of meat is both acceptable and synonymous with the use of animals as a tissue source in replacement surgery. Such *publics in general* (with the suitable acronym PIGs) are attributed public identities with whom scientific institutions often seek to be associated (Michael and Brown, 1998; Michael and Birke, 1995). In this case, the 'public in general' is the meat-eating public. We can contrast this with *publics in particular* (PIPs) comprised of specific groups which scientific institutions seek to disassociate themselves from, *particularly* the oppositional critics implied in the extract above ('you criticise').

The Imutran extract largely meshes with the dominant moral position taken by the UK regulatory and advisory institutions responsible for xenotransplantation (UKXIRA, the Minister for Health and the Home Office). Assessments of applications to conduct procedures involving experimental animals are made on the basis of a cost/benefit analysis of likely suffering to the animal weighed against potential clinical benefits. The significant point is that the axis of this balancing principle (suffering/benefit) means that the moral status of the donor animals is always negotiable vis-à-vis assessments of possible clinical value. Here the moral status of donors varies according to the species in question and assessments of the likely clinical merits of the research. Such ratios are the evaluative backbone of the *Animals (Scientific Procedures) Act* (1986). For example, in accordance with this morally hierarchical species arrangement, the UK Government has ruled out the use of higher primates for XTP whilst endorsing the moral acceptability of using pigs:

> The Home Office has responsibility for the use of animals in scientific procedures and their view on the acceptability of the use of any animal as a source will depend on a consideration of the harm of the procedure against the benefit it may bring. My right hon. Friend the Secretary of State for the Home

> Office, however, has indicated that he is not prepared to sanction the use of Great Apes for use in scientific procedures.
>
> (P. Boateng, MP, *Hansard Written Answers* for 12 Jan 1998)

The varying moral status evident here, having its basis in a sense of human specialness, clearly contrasts with the fixed view of moral equivalence drawn upon by animal advocacy NGOs:

> ... the whole notion of cost-benefit presupposes ... the flourishing of some as in some way dependent on the suffering of others ... the philosophical basis of animal rights is a human-nonhuman animal continuum, so animal rights, by necessity, includes human rights.
>
> (*Uncaged Campaigns*, report on xenotransplantation)

When dealing with the cultural terms of the xenotransplantation debate, XTP proponents must also confront responses associated with disgust, euphemistically referred to as the 'yuk factor'. Popular representations of xenotransplantation almost always play upon elements of the symbolism of body and species differences and the way in which those boundaries are threatened by transgenic technologies.[7] The dynamics of disgust in transgenics are treated more extensively elsewhere but it is worth raising because the theme points to contradictions across moral (difference) and scientific (similarity) (Brown, 1999). In transgenic and surgical hybrids such as these, translations of difference in morality and similarity in science are brought together and required to share the same embodiment: in other words '*how can we be so morally different if we're so physically similar*'.

Animal advocacy arguments, on the other hand, endeavour to reduce moral differences between humans and other animals. For this reason, it would be somewhat inconsistent for them to draw upon discourses of disgust, which are fuelled by a morally hierarchical species structure and assertions of the putative specialness of human identity. And yet there are examples of where the NGOs strategically draw upon the negative imagery of animals symbolically polluting human identity:

> The human xenotransplantation patient will become a literal chimera ... It sounds like scare-mongering, but let me assure you that the word chimera is being used by xenotransplantation scientists ...
>
> (Statement made at the launch of the joint report on xenografting by the BUAV/CIWF 1998)

As others rightly observe, intuitive feeling responses and subtle values like disgust have an important contribution to make to debates in new biologies exactly because they are both indicative of a sensitivity to actual risk and yet most susceptible to being dismissed as irrational and fleeting (Brown, 1999; Hilhorst, 1998; Chadwick, 1994).

Evidently, the upper hand in the debate over the moral status of donor animals has been claimed by both regulatory and biotechnology institutions within the dominant rubrics of a negotiable moral status for animals. For this reason, it is less surprising that the animal advocacy groups have strategically drawn upon pollution fears that are incommensurate with their underlying moral framework but which offer opportunities to mobilise oppositional sentiments.

Conclusion: science, culture and contradiction

How might we evaluate the relative positions of these constituencies in respect to their moral and scientific resources and the consequences of the contradictions in their arguments?

First, as Nelkin points out, the most recalcitrant of controversies are those between constituencies with fundamentally opposed moral principles (1995). Therefore, arguments between the disputants in the xenotransplantation controversy, like cloning and foetal research for example, are qualitatively distinct from debates which focus on technical disputes about say mobile phone safety, the location of nuclear repositories or the carcinogenicity of overhead power cables. Just as reducing the age limit of embryos in foetal research does not alter the moral opposition of anti-experimentation campaigners, the Home Office decision to exclude the use of higher primates for XTP on 'cultural grounds' did not alter the NGOs adherence to the position, that the use of *any* species is morally wrong. In cases such as this, the rhetorical space of morality is notoriously rigid, allowing antagonists few opportunities to do anything other than publicly restate a fixed moral agenda.

Yet the case illustrates a number of strategic opportunities which are available to the NGOs. One example is the contradictory deployment of the disgust discourse. But more important in creating a potentially more flexible negotiating space for the NGOs is the shift towards contesting xenotransplantation on scientific and technical grounds. Here there is potentially greater opportunity for both parties to contest one another's claims. This typifies the recourse to science by environmental and ecological organisations whose moral and cultural reasoning has failed to engage dominant regulatory and commercial institutions (Yearley, 1995). The chart below maps out the main positions taken by both constituencies and some of the main contradictions in their moral and scientific reasoning.

Scientific terms of debate

Biotech firms: *Clinical efficacy: transgenic pigs are similar enough to humans* *Contradiction: in respect to pathogenic risk, transgenic pigs and humans are too similar*	*Animal Advocacy NGOs:* *Clinical efficacy: transgenic pigs, like all animal models, are too different to humans* *Contradiction: exploiting similarity (pathogenic risk)*
Biotech firms: *Assertions of moral difference*	*Animal Advocacy NGOs:* *Assertions of moral equivalence* *Contradiction: exploiting the moral difference implied in discourses of disgust*

Cultural terms of debate

Figure 14.1 The dynamics of contradiction in the xenotransplantation debate

The STS perspectives mentioned earlier in the chapter tend to present NGO organisations as being at something of a disadvantage when they try to occupy the epistemologically privileged territory of science. The inequity arises because of asymmetrical access to the resources from which scientific prestige is derived. If NGOs want to contest xenotransplantation of scientific grounds, they do so with fewer symbolic and material resources than scientific institutions who can exercise authority in both morality and science. So much of the success of science depends on keeping politics and society close to hand but out of science itself (Latour 1993; Gieryn 1983). For example, the integrity of 'scientific neutrality' in biosafety assessments can be bracketed off, with varying success, from accusations of the influence of commercial pressures of venture capital investment. Evaluations of the Novartis report on PERVs were evaluated in the context of its position as both a commercial investor in the technology and also an arbiter in the determination of its safety. Debates on GMO field trails or BSE/CJD containment policies have similarly focused on the 'intrusion' of social and commercial factors into an otherwise insulated science. This does little other than buttress an idealised and bracketed science which can subsequently serve as a measure with which to judge erring institutions.

However, there are certain advantages which non-scientific groups have over scientific institutions even in matters of science. These advantages are related to differences in the way scientific institutions and NGOs exploit or are undermined by contradictions in scientific argument. In the context of science, both constituencies have generated contradictions. In making assurances of clinical efficacy, XTP scientists can not argue for enough host-donor similarity without arguing for too much in respect to pathogenic risk. On the other hand, the NGOs mobilise their customary argument that humans and other animals are too dissimilar for either modelling or xenotransplantation, whilst at the same time mobilising the transpecies risks of similarity. Now these contradictions are potentially disastrous for the biotechnology constituency whilst very promising for NGOs who have been able to draw upon secondary scientific sources which contradict one another.

In effect, it might be argued that exclusion from the privileged signs and symbols of science enables largely non-scientific constituencies to mobilise scientifically inconsistent positions. Such inconsistencies would be more difficult to maintain by scientific communities with a primary research capacity and the mandate to revoke contradictory empirical findings. Indeed, it is the revocability of scientific claims which makes science rather than morality a much more flexible negotiating space for the NGOs.

To some extent, whether inconsistent claims count as strategy or fatal contradiction depends on the putative values distinguishing scientific constituencies from their pressure group opponents. The epistemologically privileged strengths of the first constituency inhibit strategic flexibility because of the constraints of the institutional rules/rhetorics/values (CUDOS) from which they derive their privilege in the first place.

As such, we might conclude that there is the need to revisit Gieryn's assumption that a successfully bracketed 'science' almost always privileges its authors. In contexts like XTP, where the logic of immunological similarity leads inexorably to a logic of transpecies viral risk, the rhetorics from which bracketed scientific authority is built (revocability, neutrality, disinterestedness, consistency, etc.) reduce the scope to accommodate contradiction. For the NGOs on the other hand, because they are less bound by the rigidity of these rules, they therefore have greater scope for scientific arguments which are contradictory but which are also strategically more flexible than would be the case in a moral debate where the negotiable and nonnegotiable moral status of animals clash.

It is difficult to gauge whether and in what way animal advocacy organisations have managed to have some bearing upon the course of an evolving regulatory position on XTP. It seems almost certainly the case that they have lent force to continuing caution taken in respect to transpecies disease even if their moral agenda have proved considerably less effectual. In the context of UK regulation, promoters of xenografting technologies undoubtedly face an uphill struggle. Less than convinving preclinical trail studies have called into question the medium-term therapeutic promise of the approach and encouraged a regulatory re-evaluation of the licensing of animal procedures in the field. The now dismembered UK biotech firm, Imutran, initially anticipated conducting XTP clinical trials back in 1996, the height of the UK BSE/CJD crisis. As a consequence, trials were delayed until further research on zoonoses had become available. Recent work on alpha-gal 'knock-out' clones will however encourage a new round of US-based animal studies, although the UK is unlikely to see the same scale of research activity that characterised the late 1990s.

Much of the future of this debate depends on how these contradictions are either defused by the biotech companies or exploited by anti-XTP NGOs. The success of the NGOs probably lies in continuing to exploit the flexibility which allows them to deploy both moral and scientific arguments which are contradictory but which highlight controversy between scientific communities. For promoters of XTP, the future really lies in successfully mobilising the boundary between virology and immunology, that is, teasing apart the risks of not being similar enough and being too similar such that they are no longer debated on the same terms.

Notes

1. The statement cites the following supporting article: White, D.J.G. (1996) Transgenic pig organs: are they concordant for human transplantation? *Xeno*, 4, pp.50-54.
2. A striking iteration of this point can be found in the controversy over the ageing of Dolly, finding that her genome expressed internally varying rates of ageing. The report indicated that genetic traits in Dolly were inconsistent with her age but consistent with the animal of whom she was a clone, meaning that she was simultaneously three and nine years old at the time of the report (*Nature*, 399, 316-317).
3. *www.uncaged.co.uk/XenoFour.html*
4. These applications included: extra-corporal purfusion through pig spleens, livers and kidneys; pig to human skin grafts, the use of pancreatic pig islet cells for diabetes.
5. *New Scientist*, 4 September 1999, p.19.
6. *Sunday Times*, 5.7.92.
7. A number of authors have also explored the dynamics of disgust discourses in conventional transplantation (allografting) within the human species (Calnan and Williams, 1992; Wiebel-Fanderl, 1996).

References

Bauman, Z. (1991) *Modernity and Ambivalence*, Polity Press: Cambridge.
Beck, U. (1992) *Risk Society: Towards a new modernity*, Sage: London.

Brown, N. (1999) Xenotransplantation: normalising disgust, *Science as Culture*, 8, p. 3.

Calnan, M. and Williams, S. (1992) Images of Scientific Medicine, *Sociology of Health and Illness*, 14, 2, 233-254.

Chadwick, R. (1994) Corpses, recycling and therapeutic purpose, in R. Lee and D. Morgan (eds) *Death rites: Law and ethics at the end of life*, Routledge: London.

Cozzi, E. and White, D.J.G. (1995) The generation of transgenic pigs as potential organ donors for humans, *Nature Medicine*, 1, 964-6.

Elston, M.A. (1994) The anti-vivisection movement and the science of medicine, in J. Gabe et al., *Challenging Medicine*, Routledge: London.

Gieryn, T.F. (1983) Boundary work and the demarcation of science from non-science: Strains and interests in professional ideologies of scientists, *American Sociological Review*, 48, 781-795.

Gouldner, A.W. (1980) *The Two Marxisms*, Appendix Three – *On Social 'Contradictions'*, Oxford University Press: New York, 168-173.

Haraway, D.J. (1997) Modest *witness@second* millennium. FemaleMan meets OncomouseTM, Routledge: London.

Hilhorst, M. (1998) Xenografting as a subject for public debate, in P. Wheale, R. von Schomberg and P. Glasner, *The social management of genetic engineering*, Ashgate: Aldershot.

Hopkins, J. (1999) Study gives reassurance on safety of xenotransplantation, *British Medical Journal*, 319, 533.

Jasper, J. and Nelkin, D. (1992) *The animal rights crusade: the growth of a moral protest*, Free Press: New York.

Latour, B. (1987) *Science in Action*, Harvard University Press: Cambridge, MA.

Latour, B. (1993) *We have never been modern* (trans. C. Porter), Harvester Wheatsheaf: London.

McCurry, K.R. *et al.* (1995) Human complement regulatory proteins protect swine-to-primate cardiac xenografts from humoral injury, *Nature Medicine*, 1, 423-7.

Michael, M. and Birke, L. (1995) Animal Experiments, Scientific Uncertainty and Public Unease, *Science as Culture*, 5, 2, 248-276.

Michael, M. and Brown, N. (2000) From the representation of publics to the performance of 'lay political science', *Social Epistemology*, 14, 1, 3-19.

Nature, vol. 415, pp.103-104, January 10, 2002.

Nelkin, D. (1984) *The creation of controversies*, Norton: New York.

Nelkin, D. (ed) (1992) *Controversies: Politics of technical decisions*, Sage: Newbury Park, CA.

Nelkin, D. (1995) Science Controversies: The dynamics of public disputes in the United States, in S. Jasanoff, G. Markle, J. Perersen and T. Pinch (eds) *Handbook of Science and Technology Studies*, Sage: London.

Paradis, K. *et al.* (1999) Search for cross-species transmission of porcine endogenous retrovirus in patients treated with living pig tissue, *Science*, 285, 1236-1241.

Shapin, S. and Schaffer, S. (1985) *Leviathan and the air pump: Hobbes, Boyle and the experimental life*, Princeton University Press: Princeton, NJ.

Thompson, C. (1995) Humanised pigs hearts boost xenotransplantation, *Lancet*, 346, 766.

Virilio, P. (2000) *The Information Bomb*, Verso: London/New York.

Welsh, I. and Evans, R. (1999) Xenotransplantation, risk, regulation and surveillance: social and technological dimentions of change, *New Genetics and Society*, 18, 2/3, pp. 197-217.

Wiebel-Fanderl, O. (1996) *Life with a donor heart.* Paper presented at the International Oral History Conference, Gothenburg, Sweden, 1996.

Wynne, B. (1989) Sheep Farming After Chernoble: A Case Study in communicating scientific information, *Environment Magazine*, 31, 2, 10-15, 33-39.

Wynne, B. (1991) Knowledges in context, *Science, Technology and Human Values*, 19, 1-17.

Wynne, B. (1995) Public Understanding of Science, in S. Jasanoff, G. Markle, J. Perersen and T. Pinch (eds) *Handbook of Science and Technology Studies*, Sage: London, 361-391.

Yearley, S. (1992) Green ambivalence about science: Legal rational authority and the scientific legitimation of a social movement, *British Journal of Sociology*, 43, 511-532.

Yearley, S. (1995) The Environmental Challenge to Science Studies, in S. Jasanoff, G. Markle, J. Perersen and T. Pinch (eds) *Handbook of Science and Technology Studies*, Sage: London, 457-459.

15 Constructing the Scientific Citizen: Science and Democracy in the Biosciences

Alan Irwin

The 1990s were a very significant period for science and public policy and, especially, for science/public relations in the UK. The BSE crisis built up steadily through the decade – and remained a focus of lively debate and policy activity at the start of the new century. By the late 1990s, another science/public issue threatened to eclipse even mad cow disease: environmental and food safety concerns over genetically modified organisms (GMOs). This paper examines the construction of science-citizen relations within late-1990s discussions over the biosciences.

While the British government's handling of the BSE crisis has been much criticised, the initial response to increasing public disquiet over GM foods suggested that little had been learnt. Certainly, Prime Minister Tony Blair seemed uncharacteristically out of touch with public sentiment on the issue. As one (generally Labour-supporting) tabloid newspaper put in a February 1999 headline:

> THE PRIME MONSTER.
> Fury as Blair says: "I eat Frankenstein food and it's safe."[1]

The same front-page article – illustrated by an artistic impression of Blair as the monster – went on to report that Blair was "frustrated" that the "potential benefits of GM food are being ignored in the escalating row". The Prime Minister also revealed that he is happy to eat "Frankenstein food" – and indeed that he gives it to his children.

This last comment irresistibly reminded British readers of a former Conservative agriculture minister's public feeding of a beef burger to his daughter at an early phase in the "mad cow" crisis. The rather depressing implication seemed to be that, despite the widely held view that science/public relations had been badly managed in the BSE case, very little had actually been learned at the highest political level. Once again, an uncertain field of science was being employed as the basis for categorical assurances over safety while public concerns were arrogantly dismissed as irrational and emotional.

However, this negative assessment of the late-1990s relationship between science, policy-making and the wider publics must be tempered with a broader view of the changing context for public understanding of science in the UK. For example, in 1997 the government's chief scientific adviser produced a set of principles for government departments concerning the use and presentation of scientific advice in policy making. The emphasis here was very much on openness and consultation. This suggested a considerable change from the previously confidential treatment of scientific advice within government – and indicated the chief scientist's personal commitment to improving the quality and public credibility of such advice.

In October 1998, the Royal Commission on Environmental Pollution (RCEP) produced its influential report on *Setting Environmental Standards*.[2] This broad-ranging review advocated much greater transparency and openness within decision-making. It also stressed the significance of public engagement and participation – with particular emphasis on public trust and the articulation of environmental and social values:

> Those directly affected by an environmental matter should always have the accepted right to make their views known before a decision is taken about it. Giving them that opportunity is also likely to improve the quality of decisions; drawing on a wider pool of knowledge and understanding (lay as well as professional) can give warning of obstacles that, unless removed or avoided, would impede effective implementation of a particular decision….[3]

The RCEP report highlighted the relationship between science and uncertainty, the importance of public confidence in scientific developments and also possible mechanisms of public deliberation. The report added weight to arguments for a more democratic and open treatment of science. In so doing, it also demonstrated an awareness of recent findings from social science (for example, concerning the significance of public trust in institutions and the centrality of ethical concerns within public risk assessments).

This twin phenomenon of a newly *harmonious relationship* between UK policy processes and social scientific research, and of a much *greater degree of openness* to public evaluations, was to feature even more strongly in the 2000 report on *Science and Society* from the House of Lords Select Committee on Science and Technology.[4] As with the previous policy initiatives, the emphasis of the Lords report was on increased openness and transparency in the treatment of scientific advice, the recognition of scientific uncertainty, and the legitimacy of public values and concerns.[5] The House of Lords Select Committee began its report in the following terms:

> Society's relationship with science is in a critical phase. Science today is exciting, and full of opportunities. Yet public confidence in scientific advice to Government has been rocked by BSE; and many people are uneasy about the rapid advance of areas such as biotechnology and IT – even though for everyday purposes they take science and technology for granted. This crisis of confidence is of great importance both to British society and to British science.[6]

The Lords report identified a "new mood for dialogue". Direct engagement with the public over science-based policy making should no longer be an "optional add-on" but instead a "normal and integral part of the process". The report certainly represents a move away from the deficit theory and towards genuine changes in the cultures and constitutions of key decision-making institutions. The report specifically advocated the opening of institutional terms of reference and procedures to "more substantial influence and effective inputs from diverse groups."[7] As far as British policy debate is concerned, *Science and Society* takes us a long way from the more traditional portrayal of science/public relations as expressed, for example, in the Royal Society's 1985 report on *Public Understanding of Science*.[8]

In July 2000, the chief scientific adviser proposed a new Code of Practice for scientific advisory committees that further emphasised transparency, the need for an inclusive approach and practical mechanisms for public dialogue. It was intended also that the output from Lord Justice Phillip's inquiry into BSE and the House of Commons Science and Technology Select Committee inquiry into the scientific advisory system would feed into consultation over the new code. July 2000 also witnessed publication of the new white paper on science and innovation.[9] This document once again emphasised public dialogue although, tellingly, in a chapter focusing on "Confident Consumers".

All of this suggests that, despite the prime minister's immediate response to the GM food issue, the UK government system is undergoing a significant period of review and reassessment in terms of its handling of scientific and public concerns. As part of this process, social scientific analyses have found a more attentive policy audience than has previously been the case in the UK.

Certainly, policy calls for a recognition of the fundamental nature of scientific uncertainty, of the significance of public trust and confidence, and of the need to move beyond the deficit portrayal of public responses can all be traced back to academic research over the last decade or so.[10] It is also symptomatic of this new context for science/public relations that the published findings of the Economic and Social Research Council (ESRC) Global Environmental Change research program received a positive policy reception.

Thus, in the ESRC program's October 1999 briefing on *The Politics of GM Food*, the emphasis was on "more effective ways of handling political decisions in the face of uncertainty" and "the central need for public involvement in issues that are inherently ethical in nature rather than purely scientific."[11] For anyone who has followed science and technology policy debates in the UK over the last few decades, the congruence of official statements and social scientific findings seem remarkable indeed.

However, and as Tony Blair's remarks on GM food remind us, it is probably more appropriate to view this new mood of dialogue and public engagement as a matter of debate and contention within government rather than as an irrevocable shift. It may well be that, on balance, "dialogue theory" currently has the upper hand over deficit theory. Nevertheless, there is still relatively little UK experience (especially for government) of moving from statements of general intent to practical applications. Equally, and as the 2000 white paper emphasises, the economic pressures for continued science-based innovation are powerful within the UK.

The search is therefore on for an approach to public engagement that will permit rather than impede scientific and technological development in areas such as biotechnology and the biosciences. In such a situation, it is possible to predict likely constraints on the form and extent of public dialogue over science – especially if public discussion is seen to hinder innovation and economic competitiveness.

In this changing context, it becomes especially important to analyse the particular constructions that are being placed upon what we can term *scientific citizenship*. Does dialogue imply that public knowledges are given the same status as scientific understandings – or instead that familiar deficit notions of an uninformed public are recycled? Who for example gets to decide what counts as a legitimate problem for discussion? How are the *informative* (or information giving) and *consultative* (or information gathering) dimensions of participation to be balanced? What happens when public opinion is opposed to government policy – or, more likely, when certain shades of opinion are opposed but others are in favour? We also need to be aware of arguments concerning the "special" character of science within public discussions. As Nelkin put the general issue in the mid-1970s:

> The complexity of public decisions seems to require highly specialised and esoteric knowledge, and those who control this knowledge have considerable power. Yet democratic ideology suggests that people must be able to influence policy decisions that affect their lives.[12]

Rather than viewing science and democracy as fixed or opposing points, the argument in this chapter is that we should examine the specific configurations of these concepts – and of "scientific citizens" – within contemporary debate. Put bluntly, *how is the scientific citizen being constructed within current policy and decision processes?* This question is especially important given the apparent academic and policy need to move beyond the mere advocacy of scientific democracy and towards a more considered treatment of the possible *forms* of such democracy and their implications for the wider publics.

In order to pursue this objective, one important governmental initiative will be considered as an example of the construction of both science and public consultation. We will be especially sensitive to the framing of issues for public debate, the constitution of the "audience" for such discussions, the characterisations of science (and of scientific fact) within the initiative and the implicit model of scientific citizenship being employed.

The UK's Public Consultation on Developments in the Biosciences (PCDB) explicitly attempted an open and two-way approach to the public. Between 1997 and 1999, this government led consultation aimed to build up a public assessment of the "biosciences" (including xenotransplantation, animal and human cloning, genetic modification of food, and genetic testing). In British terms, this represented a path-breaking exercise – and one intended to have wide consequences for the operation of national regulatory policy. Announced by the minister of science, commissioned by the UK Office of Science and Technology (OST), and conducted by one of Britain's best-known market research companies (MORI, or Market & Opinion Research International), this was a high-profile and forward-looking consultation in a politically, and economically, sensitive area.

The discussion here is based upon published materials from the consultation but also the author's direct experience as consultant to the qualitative phase of the exercise.[13] This account, therefore, draws upon attendance at two advisory group meetings, extensive informal discussion with those engaged in the initiative and observation of three qualitative workshops.

1 Constructing scientific democracy

The Public Consultation on the Biosciences was an initiative without precedent in the UK in terms of the numbers of people involved in both the qualitative and quantitative research, and in terms of the focus on a range of technologies grouped under the heading 'the Biosciences'. It was thus, in effect, an experiment on a large scale.[14]

In November 1997, John Battle, Minister for Science, Energy and Industry, announced his intention to hold a public consultation exercise on bioscience issues. According to OST, Battle believed that the debate over biotechnology should be extended to include those without preconceived views. This also would allow a deeper exploration of the wider, including ethical, issues associated with developments in the biosciences. On that basis, the minister hosted a preparatory meeting with a range of interested parties in March 1998.[15] At the meeting, Battle emphasised "the significance of developments in this area and the importance of broader activity to encourage measured and inclusive debate on major scientific issues". The main purpose of the exercise was to identify and explore public hopes and concerns and to feed these into the policy process.

The preparatory meeting involved those active in the bioscience and science communication, including representatives from the Wellcome Trust, Genewatch, The Church of Scotland, the Royal Society, and the Bio Industries Association. In response to Battle's observation that there is no template for this kind of event, a range of suggestions and questions was put forward. These included (from the Church of Scotland) support for a small group and "bottom up" approach, (from Genewatch) the need to balance inputs from the natural and social sciences, and also to remedy the "dislocation between how people feel and the regulatory system", (from the European Commission) the need to have "informed" rather than "emotive" debate, (from Greenpeace) the need to recognise that judgements are informed by personal values and emotions and that these are important components of the debate.

While the preparatory meeting was broadly in favour of the new initiative, it was already clear that various parties brought contrasting models of science and democracy to the discussion. Certainly, participants placed different degrees of emphasis on the need for debate to be scientifically informative and/or citizen consultative. At the same time, there appeared to be an agreement that the consultation should ensure that "shades of grey and areas of uncertainty were explored". Equally, and as a member of the Green Alliance put it, it was necessary to have more clarity about the objective of the exercise. John Battle concluded the meeting by noting that this kind of discussion had been an experiment in itself.

Right from the start, however, it is important to set the consultation in the context of other science/public initiatives in the biosciences taking place in the late-1990s. One contemporaneous research project to have considered these issues from a public perspective was the *Uncertain World* report on genetically modified organisms, food and public attitudes in Britain prepared by the Centre for the Study of Environmental Change (CSEC) at Lancaster University.[16] Sponsored by Unilever, and with input from the Green Alliance and other NGOs, this study was based on nine focus group discussions held in the latter part of 1996. The project highlighted public ambivalence towards GMOs in food products, but noted also the general sense of inevitability and fatalism regarding such technologies. The report observed peoples "mixed feelings about the integrity and adequacy of present patterns of government regulation, and in particular about official 'scientific' assurances of safety".[17]

The Lancaster report concluded:

> Our suggestion is that the key need arising from this research is for urgent and imaginative 'institutional' experiment.... This should be aimed both at attuning industry and government better to public sensibilities, and at advancing public involvement in the crucial range of issues raised by the new commercial phase of GMO technology. The research gives grounds for concern that limitations in present arrangements, coupled to wider inadequacies in present UK regulatory culture overall, may be concealing from the view public concerns of major significance for the future.[18]

Immediately before the biosciences consultations, a second UK exercise took place – this time, with a practical and policy orientation that was even more explicit than that of the Lancaster study. *Citizen Foresight* addressed the future of the agriculture and food system.[19] As described in a report by the London Centre for Governance, Innovation and Science, and The Genetics Forum, the methodology of this initiative in democratic policy-making was designed by an expert group with experience in genetics, food, policy-making and citizen participation.

The key feature of this new exercise was the random selection of 12 British citizens who came together for 10 weekly meetings (and some 30 hours) to listen to evidence, ask questions and draw conclusions. Members of the panel then chose the particular topics for discussion. Expert witnesses appeared at the direction of the panel. In that way, and within practical constraints, witnesses could "define for themselves what they regarded as relevant expertise".[20] The members of the panel drew up their own conclusions – expressing ideas of unanimous agreement but also minority views. Ownership of the results explicitly belonged to the citizens themselves.

The main conclusions of the citizens' panel were that genetically modified crops are unnecessary, that all foods should be labelled as "GM" or "GM-free", that agriculture should be transformed away from intensive methods towards low usage of pesticides and "artificial chemicals", and that food distribution is "currently in the hands of too few supermarket companies".

Against the background of such recent initiatives, an advisory group to the biosciences consultation was appointed in June 1998. Membership was drawn from the Women's Institute, the supermarket chain Sainsbury's, the editor of *Nature*, the Green Alliance, Wellcome Trust, Zeneca, the University of East London and the Biotechnology and Biological Sciences Research Council (BBSRC). Both the constitution and status of this advisory group were matters of discussion at early meetings. For example, questions were raised about the exclusion of any organisation directly opposed to developments in the biosciences. It was also agreed that the group should have an advisory rather than steering role within the consultation since, very importantly, the OST was to be in overall control of the process. It was accepted by the group that a code of collective responsibility should apply to all decisions.

In its first meetings, the advisory group was especially keen to explore the precise meaning of the unfamiliar term "biosciences". At least one member queried the feasibility of maintaining such broad coverage since there may be significantly different public perceptions relating, for example, to food or health. Certainly the two previous initiatives had been given a much more specific focus than the biosciences. Against this suggestion, it was argued that a focus on generic issues would make the results more "applicable". This argument was reinforced by the notion that if the exercise concentrated on "principles underpinning accessibility and the use of information and the roles and remits of advisory and regulatory bodies" then it should be possible to work at the generic level.

Informal interviews within government suggest that employment of the term "biosciences" also facilitated the exercise being located in the Office of Science and Technology whereas "biotechnology", for example, might be seen as the particular responsibility of another government department. Similar questions of departmental jurisdiction were seen to apply to an explicit focus on, for example, food or health. Certainly, members of the advisory group considered that the OST presented a more neutral stance on these issues than the Department of Trade and Industry (which was characterised as a sponsor of the biotechnology industry and therefore problematic in such a consultation). At the second meeting, it was agreed that biosciences should cover "genetics research and its applications".

Immediately, therefore, we gain a sense of the institutional negotiations behind the consultation and its specific framing. Inevitably, such discussion excluded the publics whose views were considered to be central to the exercise.

The incoming Minister for Science, Lord Sainsbury, joined the third meeting of the advisory group in October 1998. By this point, a number of characteristics of the consultation clearly marked it apart from the *Uncertain World* or *Citizen Foresight* projects. As the minutes of the October meeting record it, Sainsbury expressed one of his overall priorities as being the "optimum use of scientific advice to inform decision-making, both by Government and the general population". The minister also accepted that there was a general lack of faith in the Government's use of science and in oversight processes.

> To remedy this Government has to ensure that not only are its systems appropriate, but that their existence and role are communicated. To restore public confidence in the Government's use of scientific advice required propel to understand the mechanisms used to arrive at decisions and accept that those were appropriate and based on sound principles.[21]

This suggests a decidedly loaded presentation of the science/citizen relationship and one that assumes that better communication will resolve problems of public confidence. In response, a member of the advisory group stressed that it was important to avoid the "deficit model, which simply assumed that the provision of information would ensure public confidence". Lord Sainsbury's reply was that "many basic scientific facts ought to be agreed and that the information about the regulatory system was factual, which should mean that a large amount of information…could be agreed". This drive to produce "facts" that could then unproblematically feed into the exercise was a major feature of the consultation – marking it apart from both citizen-led approaches (which aim to respond to expressed public needs) but also from sociological perspectives that suggest that the "facts" are never so unproblematic within contentious areas of debate.[22]

By this point it was also clear that the consultation was to be very much focused on the UK's system of bioscience regulation, oversight and control. As Sainsbury expressed it:

> Understanding people's knowledge and expectations of the oversight and information system was a necessary starting point to allow the identification of whether the systems themselves, or the way that their role is communicated, can be improved.[23]

Specifically, the public consultation would inform a larger governmental review of the regulatory structure for biotechnology and genetic modification. Thus, it was emphasised to the group at its fourth meeting that it was important for the new Ministerial Group on Biotechnology to be kept fully informed of developments – suggesting that the primary audience for the consultation was government itself. Chaired by the Minister for the Cabinet Office, the Ministerial Committee on Biotechnology and Genetic Modification (MISC 6) announced its intention in December 1998 to:

> …address any gaps or unnecessary overlaps in our current framework and…consider other important questions such as whether our systems could be simplified and made more transparent, and the ways in which we consider ethical and stakeholder interests.[24]

In May 1999 a new strategic structure for biotechnology was duly announced – with the public consultation cited as one specific input to this.

Meanwhile, in October 1998, Lord Sainsbury established the following general aims for the initiative:

- What is the level and nature of people's awareness of technological advances in the biosciences?
- What issues do people see arising from these developments in the biosciences and how important are these compared to other major scientific issues?
- What is the extent of people's knowledge of the oversight and regulatory process in the United Kingdom and Europe?
- What issues do people believe should be taken into account in any oversight of developments in the biosciences?
- What information should be made available to the general public from the regulatory system and about advances in the biosciences?

Sainsbury justified the initiative in the following terms:

> The consultation sets the challenging task of seeking the public's views and promoting informed debate. Our long-term aim is to encourage public confidence in the Government's use of scientific information and know-how. Understanding what people expect of Government and science is crucial to meeting their needs. I hope that the consultation will help focus the policy-making process…it already seems that the OST consultation could be a more citizen-led and participatory initiative than any carried out on science and technology in the past.[25]

Once again the encouragement of public confidence features prominently. At the same time, the intention to make the exercise "citizen-led and participatory" is spelled out clearly. Nevertheless, it is very apparent that these questions were being generated by government rather than by members of the public or even the advisory group.[26] These centrally set questions were to shape the initiative and, especially, provide the basis for public questioning.

It was agreed at this time that the initiative should collect both qualitative and quantitative evidence. As generally characterised within the advisory group, it was important that the consultation should incorporate in-depth group discussions and statistically representative individual responses. Put slightly differently, the consultation was attempting to borrow from the previous experience of qualitative and focus group research – while also preserving the "scientific" validity and generalisability of its eventual conclusions. The consultation would thereby involve far more people than had the *Uncertain World* or *Citizen Foresight* studies.

Meanwhile, the link to the regulatory review process and the consequently tight timetable put the exercise under substantial pressure from the start. While in October 1998 discussions over the consultation were still at a preliminary stage, the plan was for the whole exercise to be completed by April/May 1999 in order to inform the policy review of biotechnology regulation. The advisory group saw the possibility that the consultation could very directly inform governmental activity as a major strength of the exercise (especially when compared to the two previous initiatives discussed in this section). However, the short time-scale was also a matter of some concern to members who felt this allowed insufficient scope for discussion and consideration.

OST recommended that the recently established People's Panel should provide the sample for the initiative and that MORI, the company that runs the People's Panel, should undertake the actual consultation. The People's Panel is based on 5,000 adults selected as being "representative of the UK population in terms of age, gender, region and a wide range of other demographic indicators."[27] The Panel had been recruited in the summer of 1998. The Social Research Division within MORI managed both the Panel and the OST consultation.

Certainly, decisions needed to be made and acted upon rapidly. Equally, the consultation had now moved from its initial unformed and open stage (as announced by John Battle) to a large-scale and "representative" exercise based upon a sophisticated social research methodology.

2 The public consultation

At the advisory group's fourth meeting (also in October 1998), the two-pronged methodological approach was confirmed: qualitative discussion groups (or workshops) and a larger quantitative survey. However, when in December, and following the perceived success of the qualitative pilot study, the continued need for the quantitative survey was challenged, the justification was provided as follows by the OST chair of the advisory group: "…in order for the study to be taken seriously by ministers and other observers, a quantitative stage would be essential…" This official requirement for quantitative data and the implied down-grading of qualitative research (at least in terms of political impact) represents a significant issue for public understanding of science research – and one that deserves further discussion and comparative analysis.

In addition, the particular requirement for ministerial credibility suggests a possible tension between the citizen-led and policy-informing intentions of the exercise. The institutional pressures to speak to government in a recognisable fashion were very apparent. The specifically *scientific* focus of the consultation on the biosciences was highly relevant to this aim since again it allowed public opinion to be expressed within the operational categories of government.

The qualitative phase of the research was eventually based on two-day workshops held in six venues around the UK. This involved some 120 members of the public. The quantitative work included over 1,100 members of the People's Panel in interviews.

The workshops for the qualitative phase followed a carefully structured format. This first day covered:

- General awareness of the biosciences in the context of other areas of scientific and technological change;
- Questions of influence and trust;
- A discussion of regulation – who is, and who should be involved?; and
- Public views on information (including questions of what should be made available to the public and how trustworthy and reliable this should be).

Within separate groups (or "syndicates"), participants were asked to consider specific topics: fertility and reproduction, genetic testing/screening, gene therapy, xenotransplantation, medicines, cloning, animals and microbes, plants and microbes. As part of these discussions, showcards and handouts were used to stimulate discussion and raise important issues.[28] Handouts aimed to provide factual information, while showcards were explicitly designed to encourage public debate and discussion.

The design and content of these materials was a topic of lively debate within the advisory group – especially with regard to their scientific accuracy and accessibility to a non-specialist audience. Stimulus materials were not only read by the Advisory Group and MORI, but also by the Department of Health. An established team of science writers prepared the materials. This discussion reflected Lord Sainsbury's concern both to inform and collect public views – and, significantly, to orient discussion around officially recognised "scientific" issues.

The determination to map public views on to technically and institutionally defined issues represents a very important feature of the consultation. Viewed critically, this "pre-framing" of the agenda, as expressed in Lord Sainsbury's five questions, restricts the possibilities for public responses to operate within their own terms of reference and frameworks. In illustration of this, there was little scope for members of the public to challenge whether any comparison across "scientific issues" is valid or, for example, whether the key issue is information or political empowerment with regard to science and technology. Qualitative research into environmental matters has previously suggested that public views of pollution do not stand apart from the wider constructions of everyday life and meaning.[29]

Thus, one study of local responses to chemical hazards found that public assessments of risk were inseparable from a larger sense of social powerlessness and distrust in governmental institutions (and politicians of all kinds).[30] In such a loaded context, analytical distinctions between different scientifically defined issues (in this case, between chronic health problems and the risk of large scale explosion) become less important than the wider social and personal concerns over welfare and quality of life. While the consultation separated the biosciences into particular topics, the possibility remains that this framing misses out on more pervasive problems and anxieties. Equally, the construction of the exercise around issues likely to be unfamiliar to participants and then providing factual information to overcome their assumed ignorance, suggests a return to the deficit theory of public groups as operating in a knowledge vacuum.

This question of pre-framing the agenda represents a central issue for consultations of this kind, especially in emerging areas of scientific concern where researchers will inevitably find themselves both generating and collecting public views about topics that have not previously been considered – and doing so in an unavoidably artificial and decontextualised fashion. In the consultation's defence, and based on observation of three workshop sessions, members of the public demonstrated a high level of engagement and interest – especially given the standing start with which they began the discussions.

Certainly, members of the public were generally keen to take away information on the biosciences at the end of the first day of the workshops. Typically, they returned a week later with well-considered views (based partly on the materials provided but also on discussion with family and friends as well as careful attention to the media). Equally, the morning of the first day in each of the six workshop locations was unprompted in order to allow the wider expression of public views. It was also very clear to this observer that participants generally enjoyed their involvement – often expressing the view that all this was unfamiliar to them but that it had really made them think. The fact that public inputs might in some way inform government policy undoubtedly added a certain excitement and focus to the exercise.

Of course, these positive comments relate mainly to the qualitative phase of the exercise. Despite the careful design of the questionnaire in terms of drawing upon the qualitative stage and employing a limited number of open-ended questions both in the pilot and in the actual face-to-face interviews, interpretive flexibility is inevitably constrained in such a large-scale survey.[31] Equally, the individual and non-deliberative nature of interview responses impedes the articulation of everyday context and the expression of more pervasive concerns. However, a quantitative approach does allow engagement with a much greater number of respondents than can be accommodated within a deliberative workshop. There may therefore be a trade-off here between interpretive flexibility and volume of respondents.

Handouts and showcards for the first day of the qualitative phase covered the full range of topics covered by the consultation. A few examples of showcards can be offered from the plants area:

- **Are genes good or bad?** "Genes are present in all living organisms. They do not have moral characteristics. They are merely chemical components."
- **Is it natural?** "In a natural world, human beings would not fly in airplanes, send rockets to the moon, eat Pot Noodle...have governments, or live until they were 70, 80 or 90. So what is natural?"
- **Can the companies be trusted?** "Companies are people in work; this helps solve unemployment. Companies can make the investment and take the risks that are needed. That saves taxpayers money. They are entitled to any profit if they make agriculture more efficient."

Day 2 of the qualitative workshops followed a week later. Once modifications had taken place following the pilot exercise, this entailed each of the syndicates proposing what they considered to be an ideal mechanism for regulating their allocated topics. In that way, the day was very much focused on issues of regulation and control from a public perspective. Three questions were presented to the participants concerning their proposed regulatory mechanism: How can it ensure trust in the process of regulation? How can it deal with scientific uncertainty? And how should the new mechanism take account of public views?

After completion of this qualitative research, the quantitative stage was organised around a detailed interview format. Following piloting and various suggestions from the advisory group, the interview addressed questions of public awareness of scientific developments, the significance and meaning of biology and genes, the beneficial (or otherwise) character of such developments, the reasons why developments were taking place, and the character of the regulatory process (including questions of trustworthiness and possible information sources). Specific questions included:

- Thinking about major scientific discoveries or development, do any spring to mind?
- When I say genes, spelled g-e-n-e-s, what if anything springs to mind?
- Now thinking about biological developments again, what things if any do you think you would personally take into account if you were deciding whether a particular development was right or wrong?
- Would you say you have had too much information about biological developments and their regulation, too little or about the right amount?
- Which, if any, of the following types of people or institutions would you trust to provide you with honest and balanced information about biological developments and their regulation? And which, if any, would you not trust?

The time-scale of the operation should be re-emphasised at this point. The qualitative phase (involving a total of 123 respondents) was conducted between December 5, 1998 and February 6, 1999. Piloting of the quantitative research (among 50 respondents) commenced on February 6, 1999 and 1,109 main-stage interviews were conducted between March 13 and April 4, 1999. The research results were published in May 1999. The speed at which all this took place clearly put the exercise under tremendous strain – as discussions within the advisory group suggested.

Indeed, such was the speed of the operation that the advisory group's report – produced so as to coincide with the delivery of the main findings – noted: "At the time of writing, the Advisory Group has not seen drafts of MORI's final report, so cannot comment on the report itself. This is one of several points in the process at which the input of the Advisory group was significantly constrained by a highly compressed timetable imposed by Government requirements."[32] Perhaps of greater significance for the initiative, and as the advisory group also observed, the timetable allowed little chance for reflection on the qualitative findings before moving into the quantitative phase (although the advisory group expressed itself "content" that sufficient discussion had taken place).

Despite being written before the final report, the advisory group singled out a number of the most successful elements of the exercise. These included the use of two-day workshops rather than 2-3 hour focus groups, the use of an in-depth qualitative phase to inform the framing of the quantitative phase, the focus on policy-making and the regulatory system, the "full transparency" of the process. The advisory group also recommended that "the Government recognise this type of consultation as a necessary, productive and continuing part of developing public policy in the biosciences. Further consultations should be conducted, focusing on particular topics".

The advisory group's report repeated many of the points that had emerged in its previous meetings. Returning to the original questions posed by Lord Sainsbury, strong reservations were expressed about "assessing the level and nature of people's awareness of technological advances in the biosciences" given the broad scope of the exercise and the limited time available. Equally, the advisory group had reservations about gauging the relative importance of issues from the biosciences "compared to other major scientific issues" – and not least because of the relatively low level of awareness with which members of the public commenced their discussions. Once again, reservation was expressed about the generic term "biosciences". "A particular concern was that it brought together the medical and agricultural applications of biotechnology, and that people's reactions to these applications were likely to be quite different".

The final MORI report appeared in May 1999. Among its key findings were:

- that the public believe advances in human health represent the biggest benefit to arise from scientific developments;
- the vast majority of people (97 percent) believe it is important that there are rules and regulations to control biological developments and scientific research;
- the main issues people say should be taken into account when determining whether a biological development is right or wrong are whether people will benefit from it and whether it is safe to use; and

- the thing that people most want in relation to the biosciences is more information on the rules and regulations.

On May 21, 1999, the minister for the cabinet office presented the main results alongside the announcement of a new regulatory structure for the handling of biotechnology. The report was authored by MORI and published as three volumes – with the third consisting of 145 quantitative tables.

3 The framework of consultation

Here then we have a carefully conducted exploration of public attitudes towards the biosciences. Although the timetable was tight, the exercise was conducted skilfully and sensitively. The advisory group noted in its report "…we have every reason to commend the results of the consultation as substantial and credible. They deserve to be taken seriously by ministers, members of the public, and commentators".

As noted above, the author attended three of the qualitative sessions. A lively public discussion was observed – and especially during the second day of the workshop format. In line with previous qualitative research, public groups expressed well-developed views on these topics (despite their initial unfamiliarity) once they had been given the opportunity to ponder and discuss them both inside and outside the workshop. Members of the public also appreciated the opportunity to develop their thinking about this unfamiliar topic and for their views to be taken seriously by government.

This chapter is, however, less concerned with matters of good professional practice than with the overall structuring and framework of this exercise. On that basis, we must take special note of its institutional framing. This took a number of forms.

First, there was the construction of the "biosciences" as a generic category despite reservations among the advisory group that this might blur the issues and lead to the domination of certain public concerns over others (for example, GM foods, which became an issue of particular media attention from February 1999 onwards – just as the quantitative phase was commencing). Second, we have noted the institutional requirement both to inform members of the public about developments in the biosciences and to gather views (suggesting a deficit theory element within the exercise). Third, there was the particular orientation of the exercise towards regulatory policy and oversight – with significant consequences in terms of the necessary time scale but also for the steering of public responses. Fourth, we witnessed the perceived need for institutional credibility as provided by quantitative data rather than qualitative responses alone. Linked to this, considerable importance was attached to statistical representativeness – presumably since it would be inappropriate for public policy to be guided by the reflections of a small group of citizens (a requirement that does not seem to be uniformly applied within the exercise of democracy).

One consequence of this institutional framing was that these participants appear as essentially *reactive* members of the public rather than as citizens in any more active sense of that term. Going further, this framing led to the consultation taking shape as a highly sophisticated exercise in *social research* rather than as a citizen jury or a direct discussion between government and citizens. While there was scope within the investigation for unprompted and spontaneous responses from the public, the overall format was very much shaped by the governmental sponsors, the advisory group and the MORI researchers. It was perhaps for this reason that the House of Lords Select Committee took a very critical view of the exercise. In a short section on the consultation, this author is cited as offering the view that lay participants were "engaged by the issues" and developed "rich understandings". Nevertheless, the committee concludes: "this in itself does not justify the process… Indeed, despite its name, we see the exercise as closer to market research than to public consultation".[33]

One major advantage of the highly structured nature of this exercise is that it facilitated the construction of public responses in such a way that they can "speak to" government and therefore allow practical engagement with national issues and priorities. The disadvantage is that, despite the significant spontaneous and unplanned element within the research, public concerns were accommodated within the scientific framing of the biosciences and of one (albeit important) agenda that may or may not be shared by members of the public. It was thus the case that the format of the exercise assumed that development of the biosciences in some form would indeed occur – leaving those with fundamental objections sidelined within discussion (as one participant pointed out at the end of a workshop).

The assumption from the start was that members of the public should respond to the questions generated by government and the advisory group rather than, for example, members of the government and officials being obliged to respond to public questioning. In that sense, this exercise insulated government from public scrutiny while claiming to be participatory. In this social research framework, direct engagement between citizens and, for example, the advisory group would constitute a form of bias.

The next feature of the consultation that deserves our attention is the social construction of *audience*. Despite the early rhetoric of being "citizen-led", it is hard to resist the conclusion that the prime audience for this government initiative was government itself. From the selection of the biosciences to the particular concern with regulation, and from the formation of questions to the rigid timetable, this exercise in scientific citizenship was conducted with one particular institution in mind.

This point was very clear in the second meeting of the advisory group when one OST official stressed the high profile of the exercise since it had already been "referred to in Ministerial correspondence, select committees and Parliamentary questions".[34] Of course, the construction of this particular audience by no means negates the value of the exercise. As noted above, the novelty of "ordinary citizens" speaking to government contributed greatly to the significance of the proceedings. However, it does emphasise the very limited degree to which this consultation could be described as "citizen-led".

This observation suggests the practical significance of the *institutional location* of public consultation exercises. Inevitably, the organisation of the biosciences initiative primarily by civil servants has consequences for priority setting and direction. In this case, location in OST brought both advantages and disadvantages as we have discussed.

Linked to these points, the exercise was very much *science centred* in its orientation.[35] The first phase of the exercise was trying to educate and inform as well as simply to listen. Much discussion took place within the advisory group concerning the technical objectivity, accuracy and neutrality of the briefing materials. What seems especially noteworthy about the information provided is that, despite the advisory group's very apparent concern to maintain scientific accuracy, such statements inevitably combine social and scientific assumptions. This is very explicit in the examples offered above – and notably in the treatments of trust, the "natural" and the more neutrality of genes.

Thus, even if one agrees with the statement that companies are "entitled to any profit if they make agriculture more efficient", this can in no way be seen as a politically neutral assertion. Once again, and inevitably, social and cultural assumptions are embedded in the structure of the consultation. What seems noteworthy here is that a different approach to neutrality seems to have applied according to whether statements were construed as either factual (handouts) or non-factual (showcards). The apparent assumption within the exercise was that the "hard facts" could be separated from matters of judgement and opinion. However, the selection of what counts as hard fact represents an inevitable judgement on the part of the exercise's promoters. Furthermore, and seen from an outside perspective, this fact/value distinction can be viewed as an attempt to limit rather than enhance discussion of the core issues. Within the exercise, there was very little scope for such hard facts to be exposed to critical scrutiny or contestation by more critical social groups outside the immediate group of advisors.

These points about institutional framing extend also to the main *data analysis and presentation*. This is not a citizen-written report but rather the work of a professional organisation seeking to represent the views of citizens (a very important distinction in terms of scientific citizenship).

In one example of this, the concept of "net beneficial score" is employed within the report. This is defined as the proportion saying something is beneficial to society, minus the proportion saying it is not. Thus, the development of new medicines receives a score of +57, transplants +50, and cures for disease +42. Cloning, meanwhile, scores –55, genetically modified food –44, and genetic modification of plants and animals –27.

Issues of trust are dealt with in similar aggregated fashion. Seventy-one percent of the British public indicated that they would trust their family doctor to make decisions on their behalf in the regulation of the biological sciences.[36] Thirteen percent said they would not trust general practitioners (GPs) in this way. The net trust figure is accordingly +58. This compares well with the media (–57), industry/manufacturers (–49), retailers (–61) and environmental groups (+36).

There are certainly reasons to be sceptical about such "net" figures – and not least because of the static, decontextualised and one-dimensional treatment of both benefit and trust that is being offered. More generally, the very notion of "net" in this context suggests a broad, generalised and researcher-driven model of public attitudes.

Rather than trying to undermine a very innovative and important consultation exercise, the intention in this section has been to draw attention to its operational framework and working principles. The whole point is that assumptions concerning, on the one hand, the practical needs of policy and, on the other, the relationship between science and citizenship are unavoidable in such situations. The argument of this chapter, however, is that rather than viewing these as unfortunate weaknesses, we need to acknowledge, explore and scrutinise their character and, as necessary, open them up to larger debate and inquiry.

4 Technologies of the biosciences, technologies of community

> New 'experts of community' have been born, who not only invent, operate and market these techniques to advertising agencies, producers, political parties and pressure groups, but who have also formalised their findings into theories and concepts.[37]

This study of the intricacies of one consultative exercise may appear to have taken us far from the wider debates over science and democracy with which this paper began. However, the argument of this paper is precisely that we need to move beyond general exhortation alone over such matters and instead explore the social processes, underlying assumptions and operational principles through which scientific citizenship is constructed in particular settings.

Inevitably, presentation in this manner has suggested a rather critical view of the exercise. However, it is in the character of such practical initiatives that particular compromises over both science and democracy must be arrived at – and that these should be open to subsequent reflection. In its combination of quantitative and qualitative evidence, breadth of coverage and responsiveness to public views, this initiative contained many of the key elements that other exercises might reasonably emulate.

In very specific terms, a number of practical points have emerged during this discussion that might be useful for future practice. These include: the importance of the *institutional location* for any exercise; the balance of *information and consultation*; the extent to which the *pre-framing* of consultation agenda (as in Lord Sainsbury's original questions around which the initiative was subsequently built) can affect the form and outcome of those discussions; the degree of *activity (or passivity)* that is accorded to citizen groups; and the significance of underlying *social and technical assumptions* (for example, about the coherence of the "biosciences" in terms of public assessment and evaluation).

Other practical points relate to the lingering retention of the deficit theory within this case – so that the need to "inform" public debate was implicitly premised on public ignorance about the wider issues – and the significance attached to quantitative measures rather than to the articulate and persuasive expression of public views. We also have suggested that the attempt to separate the "hard facts" from the "matters of judgement" within such debates is inevitably a problematic process – especially since it was difficult for the citizens in this consultation to seek alternative assessments or argue back in scientific terms. Finally, questions must be raised about the feasibility of addressing such complex and unfamiliar issues in an individual interview and according to a pre-determined format. Certainly, the apparent institutional drive towards quantitative evidence needs to be tempered with a sense of the appropriateness of building complex data sets upon what may be transitory and preliminary expressions of public opinion.

In more general terms, this initiative had much to say about the importance of science but rather less about the character of modern citizenship. Citizens played an important role in this initiative and participate energetically, but their contribution was ultimately refracted through the research process. Certainly the "citizen-led and participatory" aspect of the exercise was highly restrictive and suggestive of an indirect form of citizen engagement.

Within the overall framework, and as the quotation at the beginning of this section suggests, "experts of community" became spokespeople for the wider publics.[38] Meanwhile, the active engagement of social researchers as intermediaries between government policy and wider public served to give the consultation a very particular shape and character. One conclusion from this paper is that greater attention should be given to these public intermediaries and spokespeople within public understanding of science research and practice. Of course, such attention also would need to take account of the manner in which PUS researchers are themselves also constituted as spokespeople within debates.

Going further, it can be acknowledged that at least two frameworks for the relationship between science and citizenship have been in operation within this discussion. Since all such frameworks embody working assumptions and practical compromises, we can present these as competing "technologies of community".[39]

In the first place, there is the approach described here in some detail that can be labelled a *social research* framing of science-citizen relations. Operating in a highly professional and customer-responsive mode, this has sought to achieve both depth and representativeness across the population – and to encapsulate public views in a manner that is timely, relevant and digestible to policy makers. As has been suggested, this model ties in closely with current institutional agenda and working practices – and has allowed the government's categorisation of issues to form the basis for consultation.

It does not seem too fanciful to view this as a clean, clinical, and rational procedure – especially when compared to the messiness and unpredictability of environmental protests, consumer boycotts and press criticism. Within such a model, direct contact between institutional sponsors (in this case, OST and the advisory group) and members of the public is seen to constitute a form of bias.

Second, we have briefly identified what can be termed as *deliberative democracy* model of direct discussion and engagement (represented here by the *Citizen Foresight* project). This model is shaped at least partially by (selected) citizens themselves, and is based upon more restricted but also more intensive reflections. The second model carries close affinities to other established approaches including citizens' juries and consensus conferences.[40] This framing of the science-citizen relationship grants a more active role to members of the public in defining agenda and relevance. However, and as has been suggested, the link to policy concerns and practical outcomes has often been correspondingly weak in the British context.

In drawing attention to the "deliberative" model, it must be stated that this approach also suffers from various limitations. Thus, deliberative democracy experiments are typically small scale (for example, some 12 people were involved in Citizen Foresight but 123 people were involved in the qualitative stage alone of the MORI study) and are very dependent on the particular group of citizens selected. Equally, and while such an approach might claim to be flexible and unencumbered by a predetermined agenda, constraints must inevitably exist on how issues are presented and what gets covered within any particular exercise. There also is no guarantee that many of the issues identified in the biosciences consultation might not re-occur in a deliberative format (for example, the adoption of a science/cantered approach or the steering of the exercise according to a preframed agenda).

To these two existing models can be added at least one other possibility. A *qualitative and localised* model that seeks to place public assessments within the contexts of their construction and to emphasise the relational, dynamic, and discursive character of public views and assessments. As has already been implied, this fits less easily into the operational frameworks of policy-making institutions – although it does have important policy implications in terms of the advocacy of greater contextual sensitivity and the establishment of more open and two-way knowledge relations.[41] Importantly, representativeness within such research is less a matter of statistical significance than of success in identifying structural characteristics and pervasive themes and social processes.

Rather than seeking a perfect solution to the relationship between science and democracy, such models illustrate Michael's general argument that these activities feed back to the public "visions of itself".[42] It follows that rather than viewing either academic perspective or participatory style as a fixed and unchanging commitment, it is necessary to adopt a flexible and situationally appropriate approach to all models and technologies of community.

In that way, the relationship between science and democracy should not be about the search for universal solutions and institutional fixes, but rather the development of an open and critical discussion between researchers, policy makers and citizens. It is the argument of this paper that those engaged in the public understanding of science have a potentially key role to play in informing, investigating, critiquing and challenging these processes. At this point also, the discussion of science and democracy moves from the level of sloganising to an important focus for both social scientific and practical investigation and experimentation.

Acknowledgements

The author wishes to thank the following individuals for comments and advice on the chapter: Michele Corrado, Mark Dyball, Bruce Lewenstein, Mike Michael, Tom Wakeford, and three anonymous referees.

None of these excellent people should be held responsible for any errors or dubious opinions in the text.

Notes

1 J. Hardy. "The Prime Monster", *The Mirror* 16 February 1991: 1.

2 Royal Commission on Environmental Pollution. *Setting Environmental Standards* 21st report (London: HMSO, 1998).

3 ibid., 102.

4 The specialist advisers to the Lords report were Professor John Durant (Science Museum and Imperial College) and Brian Wynne (University of Lancaster). Members of the sub-committee visited both Denmark and the United States – including sessions at the University of Copenhagen (Faculty of

Social Sciences), Technical University of Denmark (Department of Technology and Social Sciences), Harvard University (Kennedy School of Government), and American Association for the Advancement of Science.

5 House of Lords Select Committee on Science and Technology. *Science and Society* 3rd Report (London: HMSO, 2000).

6 ibid., 5.

7 ibid., 7.

8 Royal Society, *The Public Understanding of Science* (London: Royal Society, 1985).

9 Department of Trade and Industry, *Excellence and Opportunity: A Science and Innovation Policy for the 21st Century* (London: HMSO, 2000).

10 For example, A. Irwin and B. Wynne, eds., *Misunderstanding Science? The Public Reconstruction of Science and Technology* (Cambridge: Cambridge University Press, 1996); D. Layton, E. Jenkins, S. Macgill and A. Davey, *Inarticulate Science: Perspectives in the Public Understanding of Science and Some Implications for Science Education* (Driffield, E. Yorks: Studies in Education Ltd. 1993); and B.V. Lewenstein ed., *When Science Meets the Public* (Washington, DC: American Association for the Advancement of Science, 1991).

11 ESRC Global Environmental Change Programme. *The Politics of GM Food: Risk, Science and Public Trust*, Special Briefing 5 (Sussex: University of Sussex, 1999), 3.

12 D Nelkin, "The political impact of technical expertise", *Social Studies of Science 5* (1975): 37.

13 At the time of writing, minutes for the first 12 advisory group meetings were available at www.dti.gov.uk/ost/ostbusiness/puset/public.htm.

14 Public Consultation on the Biosciences, *Report of the Advisory Group to the Office of Science and Technology* (1999): 2.

15 Office of Science and Technology, Biosciences Consultation, Minutes of March 10, 1998.

16 R Grove-White, P Macnaughten, S Mayer and B Wynne. *Uncertain World: Genetically Modified Organisms, Food and Public Attitudes in Britain* (Lancaster: Lancaster University, 1997).

17 ibid., 1.

18 ibid., 31.

19 London Centre for Governance, Innovation and Science, and The Genetics Forum. *Citizen Foresight: A Tool to Enhance Democratic Policy-Making, 1: The Future of Food and Agriculture*, (March 1998). Available from The Genetics Forum, 94 White Lion Street, LONDON N1 9PF.

20 David Cope, Foreword in ibid., 5.

21 Minutes of meeting held on October 20, 1998.

22 For example, H. Collins and T.. Pinch, *The Golem: What Everyone Should Know About Science* (Cambridge: Cambridge University Press, 1993).

23 Minutes of meeting held on October 20, 1998.

24 Cabinet Office Press Notice, December17, 1998.

25 Lord Sainsbury quoted in London Centre for Governance, Innovation and Science, and The Genetics Forum op. cit., 17.

26 It is noteworthy that in the advisory group report these are referred to as "Lord Sainsbury's questions".

27 MORI, *People's Panel – results from the People's Panel*, Issue No 3 (July 1999): 4.

28 MORI, *The Public Consultation on Developments in the Biosciences, December 1998-April 1999, Vol. 1* (Department of Trade and Industry): 15.

29 See, for example, R McKechnie, "Insiders and outsiders: identifying experts on home ground" in Irwin and Wynne op. cit., 126-151.

30 A. Irwin, P. Simmons and G. Walker, "Faulty environments and risk reasoning: the local understanding of industrial hazards", *Environment and Planning* A 31 (1999): 1311-1326.

31 Certainly, the MORI researchers took great care to ensure that the qualitative phase informed the subsequent quantitative investigation. Various points from the qualitative stage were incorporated in the questionnaire design for the quantitative pilot. Open-ended questions in the pilot were then used to create pre-coded categories, create new codes, back-code into existing categories, and assist in the application of a "maximum of 5 percent other" rule. M. Corrado, Personal communication, February 2000.

32 Public Consultation on the Biosciences. *Report of the Advisory Group to the Office of Science and Technology* (1999): 1.

33 House of Lords Select Committee, op. cit., 37.

34 Minutes of meeting held on July 28, 1998.

35 A. Irwin, *Citizen Science* (London: Routledge, 1995).

36 MORI, op. cit., 7.

37 N. Rose, *Powers of Freedom: reframing political thought* (Cambridge: Cambridge University Press, 1999): 189.

38 D. Pels, *The Intellectual as Stranger: Studies in Spokespersonship* (London: Routledge, 2000).

39 Rose, op. cit.

40 S. Joss and J. Durant (eds.) *Public Participation in Science: the Role of Consensus Conferences in Europe* (London: Science Museum, 1995); See also *Science and Public Policy* special issue on public participation in science and technology (guest editor Simon Joss). 26 (5) October 1999.

41 A. Irwin, op. cit: B. Wynne, "May the sheep safely graze? A reflexive view of the expert-lay knowledge divide" in *Risk, Environment and Modernity: Towards a New Ecology* eds. S. Lash, B. Szerszynski, and B. Wynne (London: Sage, 1996): 44-83.

42 M. Michael, "Between citizen and consumer: multiplying the meanings of the public understanding of science", *Public Understanding of Science* 7 (1998): 322.

16 Conclusion

Peter Glasner

This concluding chapter will address three of the issues that have arisen in the course of this book. The first, by way of summary, centres around some of the more recent societal concerns evoked by the new genetic technologies. The second will begin an analysis of the nature of the roles that the social sciences, and social scientists, play in this area. Finally, there will be a brief discussion about some of the methodological implications for social science research, particularly focussing on the use of qualitative methods.

Concerns within society at large have mainly been articulated in the areas of healthcare and of genetically modified food and agriculture. The former has been more extensively discussed in the United States, while the latter has been of the focus of controversy, and even direct action, in Britain and the rest of Europe. The new genetic health technologies are rapidly moving from laboratory research into clinical practice through a variety of pathways including the diagnosis, monitoring, prevention and treatment of a range of single and, increasingly, multi-factorial diseases. Serious concerns have arisen with suggestions that the next steps may include improving the gene pool, eliminating familial diseases, selecting traits for offspring, and perhaps even creating a superior species. However, there continue to be a range of concerned but less hostile views expressed on issues such as population-based genetic screening, and individual and family genetic testing and counselling, around informed consent, privacy and confidentiality. There are also reservations expressed about the availability of genetic tests for the assessment of reproductive risk, and prenatal testing particularly for disorders for which there are no therapeutic interventions available, as well as for mental and other multi-factorial disorders.

As genomic science continues to discover new molecular pathways for disease, it also increasingly identifies new opportunities for down stream therapeutic intervention particularly in such inherited diseases as cystic fibrosis and haemophilia, and possibly for others such as some cancers. Through pharmacogenetics, it has begun to design drugs with the optimal shape to produce therapeutic results – the right pill for the job – while minimising adverse reactions. The impact of this commercialisation process on healthcare systems has generated some concern, especially as the promise so far exceeds the delivery of more than a handful of viable products. This concern has increased with the rapid development of stem cell research and both therapeutic and reproductive cloning.

Questions of privacy and fairness extend outside the clinical boundaries as the risk of genetic discrimination grows with the identification of each new disease susceptibility gene. This raises issues of ownership of genetic information as well as informed consent about its use in such diverse areas as insurance, employment, criminal justice, and education. It goes further to raise questions of stigmatisation and of cultural difference. There are also implications for personal identity and responsibility which link in to the broader, societally-underpinning philosophical concerns about free will and determination. These then impact on the effect of genetic information on the concept of disability.

It is now clear from recent research around the world that most members of the general population, and indeed the majority of healthcare professionals, are relatively less well informed about the new genetic technologies, and the associated ethical, legal and social implications. What is apparent is that the new information generated by human genetics research is changing biomedical research, clinical practice, and public perceptions and understanding. Concerns are now being voiced about whether health professionals possess the appropriate knowledge, skill and resources to integrate this new knowledge and its associated technologies efficiently in the diagnosis, treatment or prevention of disease. Concern about inadequate education however is not something limited solely to healthcare professionals, but extends to the regulators and policy makers, and, of course, to the wider public.

Genetically modified foods and crops provide a more immediate locus of debate particularly in Europe and the United Kingdom. Here scepticism and hostility soon replaced early acceptance of, for example, genetically modified tomato paste. This was directed at companies, regulatory authorities, farmers and even major retailers. The most recent Eurobarometer survey of public attitudes shows that strong reservations are to be found across most of the countries of the European Union. In the United States, on the other hand, there has been greater acceptance of the benefits of GM foods, more confidence in the regulatory authorities, and less public resistance. In the UK, the concerns voiced primarily by NGOs and taken up by the media have resulted in attention focussing on three areas where reservations have been perceived to be most clearly expressed. These include the instigation of crop trials by government to test whether GM crops can be safely grown; the introduction through European legislation of minimum standards of food labelling so that the public can exercise informed choice; and concern about the patenting and commercialisation of GM seeds and herbicides particularly when marketed in less developed countries.

The roles of social science in developing a better understanding of genomics and the new genetic technologies goes beyond rehearsing the societal concerns listed above. These provide a useful introductory checklist of issues that can, and indeed in many cases already have, become an agenda for social research. It can be suggested that the subject matter lends itself to the application of more quantitative methods than might normally be the case in social research, and this is addressed in more detail below. However, as with any piece of specialist research, the social researcher needs not only to grasp the technical issues of their subject matter, but also be able to address the related issues that help frame the focus of their studies. The development of the social studies of scientific knowledge has clarified and accentuated this methodological advance. In particular the theoretical impact of, among others, the interests approach, social constructivism, actor network theory, and social worlds analysis, have all contributed to better understanding the contextualised nature of scientific knowledge. The social implications of the genomics revolution are so widely spread that the traditionally more focussed approach appears less useful in providing the appropriate explanatory frameworks for study.

Equally importantly, the results of any research need to be made accessible to an untypically broad range of stakeholders including scientists, health professionals and practitioners, policy makers and regulators, and interest groups, as well as the public at large. This role is likely to be extended during the research process to facilitate communication between experts, practitioners and social groups given the many misconceptions about the implications of the new genetics that abound even in specialist circles. This suggests a degree of heightened sensitivity is required to the normal reflexive understanding of the research process, particularly in being critically aware of the need to examine received concepts in a variety of cultural milieu. Social researchers cannot fail to be engaged in the processes they study, and this becomes more significant in those areas that are most contentious such as the concerns surrounding genomic science.

Finally, the reflexive researcher will also be aware that their contribution extends beyond the merely analytic and explanatory. Social research is embedded as an *actant* alongside other stakeholders in coming to grips with the implications of the new genetics. Articulating information about the processes of social and economic change necessarily involve researchers in the developments they study. Their understandings of the past necessarily involve implications for future directions and scenarios. Sometimes these will highlight unexpected areas of concern or even conflict that may make the task of disciplined social research more difficult.

There is a degree of commonality between quantitative and qualitative methods that make both appropriate for studying the inter-relationships between society and the new genetic technologies. The most significant differences however stem from the different questions that they ask. Quantitative studies tend to emphasise those answering the question: how many? Qualitative approaches seek responses to the question: what sort or kind of? There is, of course, no single 'correct' method, although it can be argued that studying how a new technology becomes embedded in society, the 'what kind of' questions are more interesting and appropriate than the 'how many' questions. There is indeed, a relative paucity of quantitative studies on the new genetics, with the honourable exception of the continuing measurement of attitude changes in, for example, Europe through the Eurobarometer surveys. The emphasis is on generalisability and comparability between and within populations, where the objectives of the research are clearly delineated, as are the uses to which the resulting data are put. Qualitative methods tend to problematise these issues, focusing more on the complexity and tentativeness of the very facts that quantitative studies wish to measure.

Society is grappling, as suggested earlier, with the commercialisation of genetically modified foods and crops, the development of new gene based therapies and interventions, the patenting of germ-lines and genome sequences, and the application of genetics to such diverse areas as the insurance industry, criminal DNA fingerprinting, cloning and xenotransplantation. It also seems to affect people at a deeply personal level, and poses significant questions about the boundaries of Nature. This collection has attempted to raise some of these issues and debates from a social science perspective, and contribute to a better understanding of the complex inter-relationships between science and society.

Index

Note: numbers in brackets preceded by *n* are note numbers.